UNDERGROUND

INFRASTRUCTURES SERIES

edited by Geoffrey C. Bowker and Paul N. Edwards

A list of books in the series appears at the back of the book.

UNDERGROUND

THE SECRET LIFE OF VIDEOCASSETTES IN IRAN

BLAKE ATWOOD

The MIT Press
Cambridge, Massachusetts
London, England

This book was set in Bembo Book MT Pro by Westchester Publishing Services. Printed and bound in the United States of America.

Library of Congress Cataloging-in-Publication Data

Names: Atwood, Blake Robert, 1983- author.
Title: Underground : the secret life of videocassettes in Iran / Blake Atwood.
Other titles: Infrastructures series.
Description: Cambridge, Massachusetts : The MIT Press, 2021. |
    Series: Infrastructures | Includes bibliographical references and index.
Identifiers: LCCN 2020044347 | ISBN 9780262542845 (paperback)
Subjects: LCSH: Videocassettes--Iran--History--20th century. | Video
    recordings--Iran--History--20th century. | Mass media and culture--Iran--
    History--20th century.
Classification: LCC PN1992.934.I7 A88 2021 | DDC 384.55/80955--dc23
LC record available at https://lccn.loc.gov/2020044347

10   9   8   7   6   5   4   3   2   1

For Jinping Wang, who taught me to write.

CONTENTS

ACKNOWLEDGMENTS

This book was the result of lively conversations, collaborations, and presentations. Defying the writerly motifs of isolation and introspection, writing this book has pushed me to be social and to build community in ways I had never experienced before. The following pages represent more than just an assemblage of ideas; they mark my own intellectual journey as I found my way through a topic about which I truly cared. I would like to thank everyone who ventured on that journey along with me.

Above all, I wish to thank those who energetically and selflessly participated in my research. Although most of the people I interviewed as part of this project remain anonymous in the pages that follow, their voices, memories, and stories animate every word. This book wouldn't have been possible without their contributions. I am especially grateful to Amir Ezzati, who gave generously of his knowledge and time; Maani Petgar who shared stories of his video store in Tehran with me; and Niyaz Saghari whose brilliant documentary motivated and inspired me.

I am fortunate to have found a dynamic group of colleagues at the American University of Beirut (AUB). All of them have shaped this book in some way: Anaheed Al-Hardan, Bana Bashour, Greg Burris, Josh Carney, Nabil Dajani, Nadia El Cheikh, Suzanne Enzerink, May Farah, Zeina Halabi, Sari Hanafi, Maysaa Kobrosly, Rima Majed, Sara Mourad, Sylvain Perdigon, Elizabeth Saleh, Kirsten Scheid, Zeina Tarraf, and Livia Wick. My time at AUB has transformed me, and I would like to thank all of the wonderful students with whom I have had the chance to work over the last several years. I am also grateful to colleagues at other institutions who have encouraged and supported my work: Kaveh Askari, Peter Decherney, M. R. Ghanoonparvar, Michelle Langford, Farzaneh Milani, Amy Motlagh, Na'ama Pat-El, Nasrin Rahimieh, Karin Wilkins, and Claudia Yaghoobi.

I appreciate all of the wonderful friends and colleagues who contributed directly to this book by reading sections or discussing videocassettes with me. Their comments, thoughts, and questions have strengthened my

ideas: Beeta Baghoolizadeh, Claire Cooley, Datev Deravanesian, Laura
Fish, Karen Grumberg, Britta Hanson, Katie Logan, Behrooz Mahmoodi-
Bakhtiari, Kathryn Maude, Hamidreza Nassiri, Pedram Partovi, Fatemeh
Shams, and Sahba Shayani. I would especially like to thank Daniel Herbert
and Lucas Hilderbrand, who read and commented on the entire manuscript.
To receive such generous feedback from two scholars I admire so much was
one of the highlights of writing this book. Special thanks are also due to
Negar Mottahedeh, who read the complete manuscript and supported my
work in many other ways. Her constant encouragement has helped me feel
at home as a scholar of media in Iran.

I would like to thank Justin Kehoe at the MIT Press and the editors
of the *Infrastructures* series, Geoffrey C. Bowker and Paul N. Edwards, for
believing in this project from the very beginning and seeing it through to
publication. I also thank the anonymous reviewers at the MIT Press for
their insightful and encouraging feedback. I am grateful to Laura Portwood-
Stacer, whose editorial touch allowed this book to reach its full potential.
Dena Afrasiabi—a tireless friend, interlocutor, and reader—helped me find
the right tone and rhythm as I wrote the manuscript; her humor and her
way with words never cease to amaze me. I would like to thank Babak
Tabarraee for his immeasurable contributions to this book and Roja Cha-
mankar for her friendship, her research expertise, and her poetry. And thanks
to *Film History* for publishing parts of what became chapter 1 in "The Little
Devil Comes Home: Video, the State, and Amateur Cinema in Iran" (*Film
History* 30, no. 1, Spring 2018).

I have dedicated this book to Jinping Wang as a tribute to our writing partnership. What began as a case of mistaken identity on our first day
at the University of Pennsylvania has become an unlikely but enduring
friendship. Even across continents and time zones, Jinping has motivated
me during the most excruciating moments of writing and celebrated with
me during the most exhilarating. She has pushed me to be a better, clearer
writer. I am certain I would not have been able to write this book without
her constant encouragement.

I owe so many thanks to family: my parents, Dean and Robin, and sib-
lings, Seth, Jessica, Ashley, Juliana, and Tyler. They have offered a steady
supply of happiness as I wrote this book. Anthony Ferraro has been a tremendous friend to me, and I cannot begin to thank him enough for all of
his support over the years. Finally, I offer my deepest gratitude to Vrouyr
Joubanian and our wonderful life together.

INTRODUCTION: THE CURIOUS CASE OF VIDEO IN IRAN

The story of the American videocassette has been told and retold. It is the story of how one film industry tamed the beast. The earliest adopters of home video in the mid-1970s hacked the technology, pushing it to extremes in order to record, collect, and circulate movies informally.[1] A decade later, however, the American film industry was making most of its profits from the sale of movies on videocassette rather than ticket sales.[2] Hollywood used the videocassette to turn movies into commodities that could be shopped for, browsed, and procured like any other consumer product.[3] This process was neither quick nor easy. It played out over several years through court battles, shifting corporate agendas, and the proliferation of franchises like Blockbuster Video.[4] Once home video technology was subsumed under the formal operations of Hollywood distribution, the practices and moral debates that foregrounded the issue of access quickly faded into the background for most Americans. In the end, Hollywood was resilient—as it so often has been.

But the story of video is not the same everywhere, and each place had to contend with this unruly medium on its own terms. Outside of the United States, the beast was not always so easily tamed. In the global story of video, *informal* industries, labor, and practices often reigned supreme.[5] As a result, everyday users continually came up against the legal, ethical, and practical dimensions of what it meant to access movies. This was especially true in Iran. In 1983, the Ministry of Culture and Islamic Guidance (MCIG) issued a ban on the personal use of all video technology, just as a burgeoning video rental industry found its footing in the country. The ban would last for over a decade. At the very moment when much of the world experienced the rise of video as a medium, videocassettes and players could not exist legally in Iran. It would be wrong, however, to assume that the absence of a legal framework for video meant that movies on video were inaccessible, a niche hobby, or a cult interest. On the contrary, by the time the MCIG lifted the ban in 1994, an expansive informal system ensured that millions of videocassettes were already circulating efficiently and widely throughout the country.

The video ban set the stage for the incredible tale of how everyday Iranians nurtured an underground world of videocassettes. Even without a formal industry to provide them with prerecorded content on video, Iranians found their way to the movies. Or perhaps better put, they welcomed the movies into their homes. They broke the law and risked their safety to gain access to sights and sounds that would have otherwise remained out of reach. The underground distribution network that developed equipped people with a diverse stock of movies on video—everything from contemporary Hollywood action movies and European art films to popular Indian musicals and midcentury Iranian melodramas. In other words, videocassettes provided access to everything that could not be found in the state-sponsored film and television industries at the time. Video dealers with briefcases full of videocassettes moved quietly through city streets and delivered videos for rent to their customers. Selecting and watching movies from the video dealer's collection was often a family affair, as videocassettes transformed the television set from simply a source of news and information to a movie screen capable of nearly anything. Although the emerging video industry of the early 1980s had been the luxury of an elite few, by the time the ministry lifted the ban in the mid-1990s, home video technology had spread to nearly every imaginable corner of the country.

How do we account for such an astonishing development, especially given the legal and material restraints of video technology at the time? Why and how did everyday Iranians operate outside of the formal state apparatus in order to circulate movies on video? What did it mean to participate in the underground distribution of this new medium? What kinds of social, cultural, and political forces determined the movement of movies on video? The video culture that developed in the 1980s and 1990s was part and parcel of other transformations in Iran. At a time when society negotiated and renegotiated public and private life in the Islamic Republic, the circulation of videocassettes accomplished two things. First, the underground distribution network helped demarcate the ever-shifting boundaries between private and public spaces under the country's new political order. While the government viewed almost every space as public and, therefore, regulatable, people fought to maintain private lives that they shielded from the state. Within these conditions, videocassettes played an important role in clarifying which spaces were risky for carrying out activities the state did not condone, as the tapes literally moved between public and private spaces. Second, videocassettes engendered an infrastructure for media distribution and consumption in the 1980s and 1990s that existed outside the domain of state regulation. In other words, home video technology settled media access as a uniquely *informal* set of practices in Iran. If scholarship often imagines Iran as a site of unforgiving regulation, surveillance, and discipline, then much less attention has been paid to porous, slippery, and informal media practices that form networks, bundle people together, and provide leisure and entertainment where it otherwise wouldn't exist.[6]

The underground world of video teaches us as media scholars a great deal about how people build vibrant cultures *beneath* repressive institutions. Media technologies play an important role in the construction of such underworlds—but often in unexpected ways. In this story, *how* videocassettes circulated mattered just as much as the sights and sounds that they delivered. It was not just that Iranians gained access to banned movie content but also that people developed and sustained an expansive infrastructure that disseminated video. They forged routes, acquired technical knowledge, broke the law, and created rituals by passing and trading hard, plastic videocassettes. It would be easy to dismiss this underground system as a neutral channel for media distribution. But such a view would overlook the fact that it was actually a rich site upon which people constructed and enacted ambivalent subjectivities in relation to the state. The government's blanket ban on video technology did little to foreclose the medium or diminish its value. On the contrary, the state's attempt to control video forged entirely new ways to engage with it.

HISTORIES OF AN UNRULY MEDIUM

The global rise of video technology coincided with an extraordinary period in Iranian history. Three remarkable events—a revolution, the establishment of a new government, and a violent war with neighboring Iraq—unfolded in tandem and would together define the trajectory of video technology in the country. First, in early 1979, the world watched in awe as a populist revolution ousted the shah Mohammad Reza Pahlavi. For over a year prior, political unrest had challenged the monarchy, as different activist networks—including communists, Islamists, and constitutionalists—joined together in protest. Over the course of the previous decade, the shah had isolated and disenfranchised almost every social and political class in Iran with heavy-handed policies, the usurping of complete political power, and severe policing institutions. The increasingly violent protests came to a head on January 16, 1979, when the shah and his family, confronted with the inevitable, boarded a plane, never to return again. What followed was a period of uncertainty. Against all odds, the people had triumphed over the shah's police state. Anything seemed possible. It was both an exhilarating and a frightening time.

The Iranian revolution—like most revolutions—operated according to a logic of informality. It rejected the state's formal institutions, including film and broadcast media. It therefore contributed to the establishment of an informal space that would help videocassette technology thrive and also come to define it. At the center of the revolution was Ayatollah Ruhollah Khomeini, a Shi'a cleric who articulated a rhetoric of revolution that centered on anti-imperialism, socialist principles, and the political importance of Islam. It is something of a contradiction to claim that Khomeini was at the center of the revolution since he was in exile in France at the time, far away from the events. Nevertheless, his voice and his ideas spread like wildfire throughout Iran, igniting people's spirits and calling them to action. It was the proliferation of consumer media technologies and their informal protocols that made this possible. Specifically, audiocassette recorders and printers allowed sermons and speeches to circulate widely in the country. Cassette tapes of Khomeini's speeches and pamphlets with his ideas—often copies of copies—provided a counternarrative to the state-controlled radio and television stations.[7] The tapes and pamphlets showed how everyday people could mobilize consumer technologies to determine the distribution of their own media content.

The exuberant period of uncertainty following the revolution was short lived. Two and a half months after the revolution's success, a new government was born. In the interim, Ayatollah Khomeini had returned to Iran and assumed a de facto position of leadership. He set in motion a new form of governance that he called the Guardianship of the Islamic Jurist. Unlike the shah's monarchy, this was a constitutional government with a Shi'a cleric at its helm. In late March, the Iranian people gathered in polling stations to vote on a referendum to formalize Khomeini's theories of governance. According to official statistics, the referendum passed with an astonishing 98.2 percent of voters in favor—a figure still contested today. On April 1, 1979, the Islamic Republic of Iran was established with Khomeini officially in charge.

The establishment of the Islamic Republic not only ushered in political transformation; it also marked a shift in cultural and social life. The new government sought to reestablish order and consolidate its power by controlling formal media institutions, especially film and broadcast media. And yet the creation of a new government did not constitute an abrupt rupture. The new regime could not simply overhaul the country's cultural polies overnight, which meant that citizens were left in the dark about which aspects of cultural life were acceptable or even legal under the new regime. It took the new government three years to establish the MCIG, leaving home video technology partially unregulated for quite some time. The ban on video that would come in 1983 was part of the larger attempt to control media in the country. Thus, for years the circulation of videocassettes flourished as an informal practice, operating in unclear legal territory, as the government struggled to articulate, determine, and enforce its formal cultural policies.

The uneven development of media policies and institutions was on par with other changes at the time. As a new government, the Islamic Republic wanted to make sure that its citizens fell in line with its totalizing ideology, and so it exerted its power into all aspects of daily life. It tried to dictate people's religiosity, to control what they could wear, watch, and consume, and to determine how they could socialize. The state claimed all power, but its laws and policies were often unclear and sometimes even unwritten. Worse yet, the enforcement of rules proved inconsistent, with punishments grossly disproportionate to the crimes. This created an ethos of uncertainty, as citizens tried to determine the boundaries between legal and illegal, permissible and impermissible, public and private. The underground circulation of unpermitted videocassettes—an illegal process in which nearly everyone participated—helped everyday Iranians navigate the boundaries between these murky categories, even if only temporarily. In other words, because videocassettes were widespread but also illegal, consumers were doggedly aware of the risks involved in accessing movies on video and of their own precarious position in relation to the state.

The social, cultural, and political transformations that began with the establishment of the new government were confounded when Iraqi president

Saddam Hussein launched an offensive attack on Iran. On September 22, 1980, an eight-year war between Iran and Iraq began. Hussein's decision to invade Iran was a power grab. Given recent political unrest from the revolution, Iran was vulnerable, and Hussein wanted control over the Shatt al-Arab, a river near the border between the two countries that provided access to the Persian Gulf. Beyond that, Hussein took seriously Khomeini's threat to spread the Islamic Revolution to neighboring countries. In the end, the Iran-Iraq War would be the longest traditionally fought war of the twentieth century, with countless casualties on either side. What began as a war of defense for Iran—as it tried to ward off Iraqi advancements into Iranian territory—became an offensive war when Khomeini rejected a ceasefire agreement in June 1982. The urgency of war had allowed the new government to fortify its power and quell dissenting voices. By positioning the war as a holy battle between Shi'as and Sunnis, Khomeini also advocated for the relevance of Shi'a Islam as a political project. He saw these as reasonable justifications to extend the war by another six years. When it ended in August 1988, none of the prewar borders had changed, but the Islamic Republic had solidified its legitimacy. The possibility of a counterrevolution was a distant sight in the rearview mirror.

Because the war began so soon after the establishment of the new government, the story of the Islamic Republic is also the story of the Iran-Iraq War. This was certainly the case when it came to policies and practices related to media. While scholars often categorize pre- and post-revolutionary Iranian media, it was the war and not the revolution that most profoundly affected the country's mediascape. At a time when the young government sought to encourage the film industry as a propaganda organ, it also recognized the importance of state-sponsored media to the war effort. It funneled its resources into fostering a formal media culture that conformed to the war's ideological project.[8] It provided subsidies and training initiatives for films and TV segments that promoted the state's particular vision of the conflict as a sacred war and as a continuation of the revolution. Many of the most important Iranian filmmakers of the last forty years entered the profession this way. The state used the urgency of wartime to justify keeping film and broadcast media at its beck and call. State-approved media content was therefore limited.

For those reasons, the war provided the perfect breeding ground for the kind of informality that fueled the underground video distribution network.

Formal media industries were so wrapped up in propagating the state's official message that they left very little room for entertainment. Films produced in this period consisted largely of poor-quality, didactic tales, while most TV programming was relegated to covering the war effort. Mediated entertainment could have provided a needed escape from the relentless messaging about the war, but such entertainment was often not available through formal channels.

Even if the kinds of films that would have entertained audiences had been available in Iran's cinema houses, the physical conditions of war made going to the movies dangerous and unpredictable. The Iran-Iraq War was not just fought on the frontlines; both sides also regularly bombed major cities, causing sudden electricity cuts and destruction to buildings like movie theaters. Given the instability outside, people sought refuge in their homes for days and weeks at a time. Home became the center of social life—it existed outside the war and often outside the state's encroaching laws and policies—and people needed something to do there. Home video technology was the perfect fit.[9]

The informal world of video only developed in response to formal attempts to regulate it. The ban did not determine the underground so much as haunt it—always present, always lurking in the shadows. The ban is part of what makes this story so interesting. Iran was one of only a handful of countries to attempt such severe restrictions on home video technology.[10] The results of this experiment—the most extreme form of media regulation—were likely unexpected for the state, which never imagined that videocassettes would flourish under such harsh regulatory conditions.

The ban coincided with the consolidation of power in the Islamic Republic and especially the establishment of institutions like the MCIG. It crystalized the regime's views on home video technology at the time. Like many policies from this period, the state articulated the ban through the rhetoric of morality. When the head of the MCIG, Mohammad Khatami, announced the ban in May 1983, he decried video as obscene and likened it to a disease that could infect all of Iranian society. In addition to its professed desire to protect Iranians' moral sensibilities, there is also good evidence to suggest that the state was also interested in economically safeguarding state-run media, like cinema and television. Home video threatened this sanctioned media ecology by siphoning away viewers.

The ban was never codified into a single law. Throughout its decade-long life, it remained a series of policies issued by the MCIG. As a result, the ban's precise meaning never became entirely clear for citizens, nor was it stable. The ban's enforcement, too, was inconsistent. People could be arrested, fined, or issued lashings for possessing or distributing videocassettes, but often they were released with little more than a slap on the wrist. The risks that people took were real, however. Many people who participated in the underground circulation of videocassettes received harsh punishments for violating the ban.

When the ban was lifted in 1994, many individual users never even noticed the shift. Until analog video technology fell out of favor in the early 2000s when digital formats replaced it, the state continued to try to limit the conditions under which people accessed the video market. After 1994, the MCIG authorized only one distribution company to deal in movies on video, and it issued permits to a meager number of titles. The underground distribution network, therefore, remained the primary means through which Iranians accessed cinema. The underground systems and processes that the ban had created continued to matter in Iran for decades to come.

Against the uncertainty of the revolution, the new government, and the war, videocassettes became a solid fixture of people's everyday lives. By the time the MCIG issued its video ban in 1983, video was already well on its way to becoming central to Iranian homes. The circulation and consumption of videocassettes provided a reliable constancy at a time when people's lives were otherwise ripped apart by political violence. Over the next two decades, Iranians would defy state mandates to consume millions of movies on video. Videocassettes as material objects did what media do best: they *mediated* the central concerns of their time and place. As I will show in detail in this book, the underground distribution and consumption of videocassettes *as material objects* provided a means of negotiating the power of the state and the agency of its citizens.

A METHOD FOR THE UNDERGROUND

To study the history of the videocassette in its social and cultural context requires that we excavate the underground. The word "underground" invokes a kaleidoscope of descriptors and images. In our cultural imaginary, the underground is—at its worst—dark, isolated, claustrophobic, and criminal. It is what Fyodor Dostoevsky famously called a "mousehole" in

Notes from the Underground: a place for outcasts and derelicts.[11] At its best, the underground fosters a world of thriving subcultures. Still the underbelly, this is where voguing, graffiti, and all-night parties become possible—a site that forges alternative visions of life, where prevalent notions of gender, sexuality, and class don't necessarily apply. As I argue throughout this book, the circulation of videocassettes discursively produced the underground in Iran in the 1980s and 1990s. In line with how we typically understand the underground, it was a space where alternative visions of national life were possible, where people negotiated what was formal and informal, legal and illegal, public and private. At the same time, unlike our prevailing conceptions of the underground, Iran's underground was not just the purview of a marginal few. It was neither dark nor did it *feel* criminal. On the contrary, the underground world of videocassettes was part of quotidian life for many Iranians of all different social classes. Perhaps counterintuitively, the underground in Iran mainstreamed informal media consumption as a facet of everyday life.

My argument in this book builds off of Rosalind Williams's assessment that the underground is a place where technologies abound. Equipped with examples like subway systems, mining sites, and telephone cables, she shows how literary works have represented the underground in order to navigate between technological pasts and futures.[12] In Iran, too, the underground in the 1980s and 1990s had a distinctly technological quality to it. It was a world produced through home video technology. Unlike Williams, I am not explicitly interested in imaginative works or the technological underground as a literary or filmic trope. Instead, the aim of this book is to explore the underground as a practical, lived experience for many Iranians in the 1980s and 1990s and to untangle the ways in which videocassette technology molded that experience.

The video underground was both discursive and material. It was shaped by state policies and people's attitudes toward them, but it was also composed of plastic shells and magnetic ribbon. Just as Karen Barad explains discursive practices as "material (re)configurations of the world,"[13] the plastic bulk of videocassettes shaped the experience of the underground and determined what it could be.[14] The materiality of video in this period shaped how users consumed, distributed, and archived movies and governed how they inhabited the underground. In this underworld, media were not just modes of transmission; they were the *objects* of transmission, too. This duality is what made videocassettes so dangerous and at the same time so precious. People became attached not just to the content that circulated but also to the material objects themselves. Videocassettes had to be secretly transported in suitcases, under trench coats, and inside laundry bundles. The video dealers who did this work always ran the risk of being caught. Users understood the value of the video precisely because of the perils that were born of its bulk.

Videocassettes bore the traces of the love affair that Iranians had with the movies at the time. Repeated views, constant copying, and overhandling wore down the material of the videocassette, especially its delicate magnetic ribbon. Lucas Hilderbrand has argued that videocassettes have captured a history of use and access in their technical failures.[15] Indeed, the underground itself was etched into the glitches, blips, and blurs often embedded in the movies on video. The videocassette embodied the underground and continues to do so today. Even decades after the ban, videocassettes still matter. Hidden under beds, shoved behind video store counters, stacked in the back of vintage shops, and piled up in storage units, they have returned to the underground. No longer watched or watchable, they endure perpetually in their boxy plastic form—reminders of a time when it wasn't so easy to access the movies. Ultimately, the physical matter of videocassettes has allowed them to matter culturally, too.

The underground in Iran is not just one place but many. In Persian, the word *zirzamini* (underground) has come to signify a range of activities and beliefs that occur outside of the state's legal and ideological framework— from the circulation of media to drugs and alcohol to oppositional political movements. At times, the various parts of the underground intersected on the material routes that supplied both alcohol and videocassettes, for example, or in the circulation of offset copies of banned books that sustained anti-regime political movements. But these connections between different components are not what made the underground legible to everyday Iranians. Despite the diversity of activities that occurred there, the underground existed coherently as a palpable crosscurrent to the regime's logic of power and control, which had otherwise saturated public life in the country. In this sense, the underground was political; it always existed in opposition to the state.

In this book, I advance two claims with respect to the underground in Iran. First, the video distribution network as a case study affords us an entry point to uncovering the inner workings of the broader underground that developed in Iran after the revolution. The narrow focus on video provides an opportunity for rich engagement with the dynamics of the Iranian underground writ large. Second, and perhaps more ambitiously, I argue that videocassettes actually helped to institutionalize the underground within the Islamic Republic. The video network may have been informal, but its operations nevertheless depended on rigorous systems, routes, and procedures. These processes structured the underground not as an alternative to national belonging but rather as a central fact of everyday life for many Iranians. The underground existed well before the advent of videocassettes,[16] but home video technology helped produce the underground as a coherent space for the many Iranians who participated in the circulation of movies on video in the 1980s and 1990s.

But how to study such a place? Even though the underground world of video and its centrality to everyday life are common knowledge to the people who lived through the ban, there is little formal documentation to show what that world actually looked like. Scholars have barely scratched the surface when it comes to the history of the videocassette in Iran.[17] There may be several reasons for this lacuna, but the most important is the methodological challenge that the underground presents. As I hope to have made clear by this point, the distribution of movies on video, although widespread, was an informal practice that operated on the periphery of what was considered legal. We know that around the globe, most people access cinema through informal means of distribution, including analog and digital video. It is only recently, however, that scholars such as Ramon Lobato have begun to develop the methods and vocabulary for studying consumer technologies and the informal distribution practices they enable.

Media historians, in particular, have struggled to account for these kinds of informal practices in their work. The underground world of video that I present in this book did not leave behind many of the traditional sources on which historians typically depend. This was an informal industry; there were no budgets, internal memos, contracts, or official correspondence. The kinds of popular sources that historians have used as a window into the everyday lived experiences of media technologies—such as magazines, newspaper editorials, television shows, and movies—also bear serious limitations when it comes to the history of video in Iran. Given the stringent regulation of media, including the press, sources that do reference video reinforce the state's stance at the time. Official sources represent video as a contraband technology and as a vessel for Western imperialism. While they acknowledge that an illicit market for video rental existed, they do little to instruct us on the mechanics, economic structures, labor schemes, and affective regimes that encouraged this underground system.

The case of home video technology in Iran thus poses an important methodological question. How do we study the history of informal media industries when their very informality resists documentation? How do we begin to map the robust underground network of video rental in Iran in the 1980s and 1990s and to study how people engaged with it? I propose oral history as a methodology to grapple with untraceable, undocumented historical media phenomena.

Much like media themselves, oral history is an inherently unstable category. It means many things to many different people. For me, oral history involves gathering historical data through sustained, open-form interviews with people who lived through, participated in, or witnessed certain events or periods. The key here is what I mean by *historical data*. Given that memories often fail us, oral history interviews do not yield facts and figures—or even some form of definitive truth. Instead, they provide narratives and stories. Those stories reveal much about the people who tell them, about their shifting values and priorities, and about what it means to remember, interpret, and reinterpret past experiences. Scholarship that draws on oral history interviews must prioritize those subjectivities rather than some illusive objectivity.

This book is thus, in part, an oral history of the videocassette in Iran. Between 2015 and 2018, I conducted forty-two in-depth interviews with a variety of people who participated in the circulation of videocassettes between the late 1970s and early 2000s. I conducted these interviews primarily in Iran, although several were with Iranian expatriates who live in the United States. While a handful of interviews were conducted in English, most of them took place in Persian. Thus, one of my roles in the writing of this book was that of translator. Sometimes that meant translating language from Persian to English, but more often it meant translating the experiences and cultural baggage that come with any language.

My interlocutors included consumers of home video technology, video distributors, film critics, government employees, video store owners, and filmmakers. The majority of people I interviewed were in Tehran when they engaged with video technology. However, I also interviewed people who were far from the country's political center during this period, including the cities of Mashhad, Bushehr, and Rasht. This may seem like a limited sample, and by some measures it is. And yet oral history is never about the quantity of interviews conducted. Each and every oral history interview opens up a world of possibilities when it comes to rethinking history and theorizing media more generally. As such, I do not mean to suggest that this book is *the* history of video technology in Iran. Rather, this book presents *a* history of video. A different corpus of interviews may have yielded an entirely different project altogether.

Researching the underground world of videocassettes is potentially risky business. Although decades in the past, this episode in Iranian history remains politically charged. The people whose stories inspired this book shared their memories of crime, war, intimacy, adolescence, and quotidian life. My oral history interviews thus became an exercise in trust as much as anything else. For this reason, my first interviews were with people I already knew—friends, acquaintances, and colleagues who had previously mentioned their relationship to videocassettes. It was through their enthusiasm for the project and their social and professional networks that I gained access to a much wider swath of video consumers and distributors, especially those residing in Iran. My command of Persian, which I began studying as an undergraduate, helped me build trust during these interviews. It also occasionally worked as a hindrance, as an American in Tehran who spoke Persian fluently could raise suspicion—and rightfully so, given the long history of American imperialism and interference in the region.

At the outset of every interview, I offered my interlocutors the promise of anonymity. Many insisted at the beginning that they didn't mind if I revealed their identities. By the end, however, most had changed their minds entirely, asking that I not identify them in my writing. This about-face suggests a lot about the oral history interview and also the underground. In excavating their memories, my interlocutors recalled the dangers of accessing movies on video; they remembered how political the video infrastructure had actually been. But more than that, during the course of our discussions, people revealed more about themselves than they had expected—about their fears, their transgressions, and their desires. Indeed, oral history as a method is important to this study because it surfaces the underground infrastructure as a site where the political and the personal penetrated the materials, processes, and technologies that fueled it.

Throughout this book, I understand the *underground* not just as a set of practices related to the informal circulation of media but also as a methodology unto itself. In an underground methodology, oral history interviews contextualize information from archival sources and vice versa. As I have noted, oral history is a central component of my research. Yet it is hardly the only source material I consider. Because the underground of Iran in the 1980s and 1990s was a site of myriad contradictions, a careful mapping requires that I draw on a multitude of sources and perspectives. In particular, archival material, including popular and legal sources, preserved the state's official stance on home video technology, even as it transformed over time. These sources—including laws, policies, newspapers, trade publications, films, and essays—were formally published, a stark contrast to my oral history interviews as discursive sources. Activating the underground as method has meant holding these two kinds of sources—oral history and archival—in constant tension, allowing neither to overpower the other. While oral history interviews attuned me to the everyday sensations of the underground, archival material often drew my attention to the state's power, which was omnipresent, even if not absolute. By bringing this divergent source material together, the underground becomes a site in which a number of in-betweens become visible, for perhaps the first time.

As one of my interlocutors, Farid, would tell me, "During the 1980s, we were like a huge colony of ants." He laughed, "We all lived underground. Burrowing and tunneling, we created an entire world that had nothing to do with up above." In some respects, the underground was literal. The war forced families into their basements, where they would wait out bombings under the surface of the city, sometimes for days on end. But the underground world Farid described meant something more than just a subterranean space that provided shelter from war. It was also a systematic and orderly way of living outside of state laws and regulations. In the chapters that follow, we will peek inside those tunnels and consider all the ways in which they were reinforced with and constituted by the hard plastic shell and magnetic ribbon of a videocassette. Such a view is not always easy. As Farid asked me, "Have you ever seen an anthill blow away with a single gust of wind? We knew that our underground world could collapse in an instant." My approach in this book attempts to document and make sense of something that was both enduringly solid and exquisitely tenuous.

WHY VIDEO, WHY IRAN, WHY NOW?

Beyond the tenuousness of the underground itself, this book also asks that media scholars take seriously—or at least notice—the very details of media access that we as users have learned to overlook. The secret life of video-cassettes in Iran reveals an entirely different vision of Iranian cinema than what scholars have imagined so far. In such a vision, cinema isn't something so much produced as it is accessed. During the course of my interviews, my interlocutors often recounted how they would watch a movie on video from start to finish, even if the images and sounds were distorted beyond comprehension. Iranians were acutely aware of the work and risks that went into accessing every movie on video. This sentiment, which I heard time and time again, tells us that sometimes *how* people access movies is more important than *what* they actually see on screen. And yet questions of distribution and access are often sidelined in film and media studies—a field that to date has favored accounts of media production and reception.[18] A recent turn to new materialism has inspired works by scholars such as Caetlin Benson-Allott, Anna McCarthy, and Lisa Parks, who have begun treating modes of access not simply as a matter of fact but rather as integral to the experiences of consuming and understanding media.[19]

What it means to access media is particularly fraught when it comes to the Islamic Republic of Iran, where reports of censorship, filtering, and jamming abound. Certainly, it is tempting to view Iran as a place where stringent, oppressive laws dictate every aspect of people's lives—and to some extent that may be true. But the very presence of an underground world of video signaled the *failure* of state policy to regulate media more than its successes. Peter Decherney has shown that regulatory policies offer a rich entry into studying the "power, politics, and art" of media.[20] Such an approach does not necessarily presuppose that government and corporate regulations are absolute. Indeed, in Iran in the 1980s and 1990s, it was the absence of effective laws and policies that allowed ordinary people to develop norms and practices that facilitated the circulation of and access to movies.[21] Embedded in those norms and practices—and even in the video-cassettes themselves—was the triumph of everyday people over the state, and even the triumph of cinema itself.

It is important that we inspect such extralegal practices. Scholarship and popular media often represent Iran and other countries in the Middle East simply in terms of their autocratic regimes. Such a framing of the region suggests that ordinary people have very little opportunity to live their lives outside of repressive laws or to reject the legal dictates of the day. In contrast, the underground distribution of videocassettes in Iran exposes the perforation of political power. Attention to this underground network highlights how the young Iranian government struggled to maintain control when confronted with the passions and ambitions of its citizens. In this regard, *Underground* joins a growing body of work that seeks to decenter state power in Iran and restore agency to the various actors who push back against, negotiate with, or refuse to acknowledge the state's regulations.[22]

The underground video network in Iran also provides an opportunity to rethink informal media beyond dominant capitalistic frameworks. An impressive body of scholarship has begun taking unregulated media practices seriously.[23] Shifting the conversation away from the ethics of such practices, these studies situate piracy as a practical means of media access for people around the world. Central to many of these accounts is piracy's relationship to copyright law, which is inextricable from capitalist logic.[24] After all, intellectual property regimes are themselves part of today's neoliberal order, treating creativity as capital and protecting large multinational corporations more than individual artists.[25] Barbara Klinger's important study of the informal circulation of *Titanic* (1999) in Afghanistan demonstrates the entanglements between piracy and capitalism. Critical reception in the United States viewed the piracy of *Titanic* in Afghanistan as a double-edged sword. On the one hand, critics decried the piracy of a globally successful film like *Titanic*, which amounted to a loss of billions of dollars for the American film industry. On the other hand, they celebrated Hollywood's ability to represent Western, liberal democratic ideals for audiences in Afghanistan.[26] Thus, Klinger's study shows how the concept of piracy is burdened with the capitalistic ambitions of copyright law.

As we will see, piracy and copyright are inadequate frames for analyzing the underground circulation of videocassettes in Iran. As Ramon Lobato suggests, "the term piracy entrenches a master paradigm—intellectual property—that should instead be dismantled."[27] Part of undoing such a paradigm involves acknowledging that intellectual property regimes do not extend to every place with equal force. Although the Iranian video network may have looked like piracy, it was structured by a different set of priorities, logics, and industrial and regulatory frames. Indeed, intellectual property would not become a serious talking point in Iran until the early 2000s, after the decline of home video technology.[28] Today Iran still is not a signatory to any international copyright agreement. While Iranians in the 1980s and 1990s recognized underground video culture as illegal, its illegality had little to do with copyright violations or commodity culture. Instead, the various informal practices that determined underground video culture resisted the state's logic of control and surveillance.

For this reason, I have avoided the word "piracy" throughout the book where possible. To impose the rhetoric of piracy onto the video circuits that animated the Iranian underground would be to evoke questions of intellectual property that are only peripheral to the story of videocassettes in the country. Perhaps worse, to evaluate the informal media practices in Iran through the rubric of copyright would be to foreclose many of the unique qualities of the underground infrastructure. As I will show, the underground video infrastructure at times shared striking similarities with informal economies elsewhere, even in those places where copyright mattered a great deal. And yet informal media practices are always unique to a particular place and time, as is the rhetoric that people used to discuss them. The Iranians I spoke with never used the word "piracy" to describe the underground world of videocassettes and only rarely used the word qāchāq (smuggling), the government's preferred word to describe the movement of both illegal media and drugs. In my interviews I more often heard the word zirzamini (underground) to describe the distribution and consumption of videocassettes. For Iranians, this word powerfully captured the multitude of informal practices and legal forces that shaped video culture at the time.

By remaining sensitive to how people describe the underground network, I treat video as a discursive object that indexes larger cultural phenomena in Iran. I see this approach as an extension of Michael Z. Newman's work on video in the United States. Newman traces semantic shifts in the word "video" since its first appearance in the 1930s to show that any given medium is not just an assemblage of technologies, practices, and aesthetic forms, but also "a shifting constellation of ideas in popular imagination, including ideas about value, authenticity, and legitimacy."[29] In Iran, too, the beliefs and values associated with video shaped it as an underground medium. While I do not adopt the same long view as Newman, one of the goals of Underground is to investigate the ways in which various stakeholders—including governmental actors, informal laborers, and everyday consumers—understood, negotiated, and contested the meaning of video throughout the 1980s and 1990s. As Brian Larkin reminds us, "the meanings attached to technologies, their technical functions, and the social uses to which they are put are not an inevitable consequence but something worked out over time in the context of considerable cultural debate."[30] In this book, I do not take for granted the inevitability of underground video culture in Iran. Rather, I examine how the informal video infrastructure became socially realized. I take stock of the material that constituted it; the ideas that fashioned it; and the political conditions under which it operated.

This book shows that, even in an age of social media and digital platforms, the legacy of home video technology continues to matter in Iran. Although the heyday of this analog medium seems decades behind us, we can learn a lot from studying its history and lasting relevance. After all, video precipitated the young Islamic Republic's first great media crisis. How the state attempted to control this unruly medium and how everyday people responded to those attempts set important precedents when it comes to contemporary media regulation, distribution, and consumption. For this reason, each chapter in this book ends not in the past but in the present. More episodic than linear, this narrative strategy is about corollaries rather than causality. In other words, I do not mean to trace an uncritical path from the past to the present but rather show that the parameters drawn by home video technology during the 1980s and 1990s still condition what media in Iran are allowed to be and what they are allowed to do, even today.

For many former video users, nostalgia is the lens through which we might understand this uneasy relationship between past and present. Like elsewhere in the world, popular media in Iran have recently contended with the legacy of videocassettes. Over the last several years, a number of essays, movies, television shows, novels, magazine issues, and social media accounts have revisited this complicated period in Iranian history. Often these representations of underground video culture betray a longing for a time when technologies and life seemed simpler and more pleasurable. This nostalgic return is not a coincidence, nor should we brush it aside as meaningless. Nostalgia is not static affect but rather a dynamic structure through which people understand the present through the past.

Thus, I claim nostalgia as both mood and method for this book.[31] By mood, I mean nostalgia articulated as a disposition, a way of looking at the world. It orients us to a certain set of questions and assumptions. Rather than deny the nostalgic hues of my sources, especially the oral history interviews, I found value in replicating and redescribing them. The publication of this book constitutes another episode in the ongoing project of nostalgia for video in Iran and beyond. But I see no problem or contradiction in that fact. As a method, nostalgia tells us just as much about the present as it does about the past.[32] Nostalgia is a generative mode rather than a valley or a lull. Indeed, it is a call to reflect on the past in order to understand the present and intervene in the future. In the pages that follow, I dive deep into the underground world of videocassettes in Iran and endeavor to tell the extraordinary tale of a time when movies were hard plastic bricks: a careful intermingling of magnetic ribbon, stainless steel screws, polypropylene shells, knobs, and reels. This was the world of home video—of videocassettes and players, of video dealers and consumers—where the underground was a way of life.

BANNED: VIDEO GOES UNDERGROUND

Nahid and her coworker Behrooz loved to talk cinema. They had a hankering for the classics, and not just Iranian movies but the greats of American and European cinema, too. It was the early 1990s, and they worked together at a government office producing short segments for state-run television. Despite their work as government employees, they spent their short lunch breaks discussing films like *Casablanca* (1942) and *The Bicycle Thief* (1948), getting lost in other worlds and imagining what it must have been like to be alive in a different time and place. It didn't matter that they were representatives of the country's formal media industry. Like everyone else they knew, they got their movies underground from a *filmi* or video dealer, who would come to their apartments once a week with a briefcase full of videocassettes to rent. Unlike most people, however, they also had small video collections of their own. They paid a lot of money to own their favorite films so that they could watch them again and again. For cinephiles like them, there was a certain pride in being able to say that they owned *Annie Hall* or *Ben-Hur* or even some obscure Japanese classic. Naturally, Nahid and Behrooz would lend each other movies from their own collections, borrowing cassettes and then trading notes a few days later.

A copy of Roman Polanski's *Frantic* (1988) was one of Nahid's most prized possessions. Not only did she love the film but, unlike other videotapes in her collection, this one was of particularly high quality. With crisp sound and brilliant colors, it was almost like an original or a mother cassette (*navār-e mādar*) as they called them at the time. Of course, it wasn't an original. In fact, this particular videocassette contained not only all of *Frantic* but also the first few salacious scenes of Alan Parker's *Angel Heart* (1987). Videocassettes were such valuable commodities that it was a shame to waste even a few minutes of the tape's full potential. While collectors like Nahid might maximize their collections by splitting films over two cassettes, video dealers often filled the remaining minutes of a tape with music videos so that their clients felt like they were getting the most value for their rental fee. This is how countless Iranians mastered Michael Jackson's moonwalk and fell in love with the sweet beats of Googoosh, Iran's premier diva in the 1970s.

When Behrooz confessed that he hadn't seen *Frantic*, Nahid, with some apprehension, agreed to let him borrow it. It wasn't that she didn't trust Behrooz to return it; rather she was always cautious about watching her high-quality tapes too many times. She knew that it was only a matter of time before the tape's magnetic ribbon began to break down or a faulty player mangled its precious contents. But after talking up the film so much, she couldn't very well deny her enthusiastic friend of Polanski's masterpiece. Nahid finally gave in and brought her copy of *Frantic* to the office one day. She told Behrooz to watch it and let her know what he thought.

Then, weeks went by without a trace of Behrooz. Nahid and Behrooz did not necessarily see each other every day or even every week, but as weeks turned to months, Nahid began to worry about her videocassette, fearing that it had met its untimely death in Behrooz's outdated VCR. She figured he was too afraid to face her. Finally, after three months without a word, Behrooz showed up with another copy of *Frantic*. He explained that the day that Nahid gave him the videocassette, guards decided to search him as he was leaving the building. They found the cassette and, when they looked through it, discovered the sex scenes in *Angel Heart*. They questioned him repeatedly over the course of two months. Even still, Behrooz was too embarrassed to seek out Nahid until he had found a replacement for her copy of *Frantic*.

That night, Nahid slid her favorite movie into her VCR to nurse her guilt for not telling Behrooz about the nude scenes from *Angel Heart* on the tape she had lent him. As *Frantic*'s opening credits rushed down the highway like cars, Nahid thought to herself, "The quality of this one isn't nearly as good as mine was." She couldn't help but laugh. Perhaps it was the sense of paranoia in Polanski's film that appealed to her, she speculated. It felt familiar. Hers was a world where movie culture existed as hurried exchanges in the stairwells of a government building.

Nearly a decade into the ban, and videocassettes were flourishing underground. They found their way to unexpected nooks and crannies, including the government institutions that sought to regulate them. After all, Nahid and Behrooz were government employees who traded their illegal stock of videocassettes at the office. Their story demonstrates that informal video distribution was central to life even at this formal, state-run media agency. Friendships, conversations, and interrogations all took place around movies on video, at least for Nahid and Behrooz. But how do we make sense of such a contradiction? Why didn't the ban prevent the circulation of videocassettes among government officials and within governmental offices, of all places? What did it mean for informal media systems to operate within formal media institutions? These questions are a useful starting point for thinking about the material and legal conditions that allowed video culture to thrive underground.

The underground video infrastructure in Iran emerged in response to a number of factors, including state policy and the unique qualities of videocassette technology. While much of the scholarship on media in the Islamic Republic has focused on formal institutions,[1] the case of video in Iran offers an opportunity to explore the ways in which informal media develop. The underground video infrastructure was a site where formal and informal practices came together, often in unexpected ways. This pattern is most evident in how the video ban—a formal regulatory policy—incited the informal world of video. Entanglements like this show that even as video operated in opposition to the formal media economy, it was never entirely separate from it. Embedded in the story of video's descent underground is an important lesson in media history: the more that states and industries regulate, the more that spaces emerge for *un*regulation—for the informal practices that are essential to what media accomplish around the world.

AN UNRULY MEDIUM

In a matter of four years, between 1979 and 1983, amateur interest in videocassettes matured into a profitable industry. During this early period, it became clear that home video technology would not be contained easily. Even as subsequent governmental institutions sought to oversee the growth of the video rental industry through increasingly stringent regulatory procedures, there remained room for people to operate outside of state control. At a relatively affordable price, the consumer technology allowed anyone with the means and knowledge to duplicate and circulate movies on video. With the potential for millions of cassettes to flood the market, the state had no way to manage how individual people used home video technology and to what end. What was certain was that the growing interest in video-cassette technology competed with television and cinema. These state-run media transmitted the Islamic Republic's ideology, rallied support for the war, and required heavy subsidies from the government. This early period between 1979 and 1983 revealed video as an unruly medium—one whose qualities threatened the very premises upon which the Islamic Republic had been built.

It would take several years for video to assume its specific cultural values in Iran. Like other places in the world, home video technology grew rapidly, especially among upper- and middle-class city dwellers. What began as a luxury item carefully packed in the suitcases of travelers from abroad soon became a budding formal industry in its own right. The mid- to late-1970s were marked by experimentation as technology enthusiasts and the country's wealthiest residents tinkered with and proudly displayed their Betamax video players. Alireza was one of the early adopters of home video technology in Iran. He first encountered Sony's Betamax player when visiting his brother, then a master's student in Los Angeles in the mid-1970s. When Alireza went to California in 1976, he saw advertisements for video players everywhere and the device intrigued him. "Watch Whatever Whenever," the famous Betamax advertisement promised. It was a tantalizing idea. A former engineering student who still enjoyed taking things apart, Alireza saved his money and persuaded his brother to bring a colossal Betamax player with him to Tehran in late 1977, before the Japanese product had even been released in the United Kingdom. For Alireza, it was an object of marvel. Its anatomy—the interlocking reels, spools, and levers—kept him entertained for hours. He admitted, however, that in terms of functionality the VCR was useless to him. Its intended purpose was to record television programs for later viewing. According to Alireza, this was not a very valuable service in Iran, where only four television channels ran intermittently in the 1970s.

Video players became more useful beginning in 1979, as the video market in Iran changed to include prerecorded content. Suddenly, the video player wasn't just a means of shifting time; it became a way to consume new content. The video industry in Iran experienced its legal heyday between 1979 when prerecorded content began flooding the market and 1983 when the state finally implemented its wholesale ban on home video technology. During this four-year period, the industry underwent tremendous growth, despite a pattern of increased regulation as governmental control shuffled between agencies. Immediately after the revolution, the circulation of video-cassettes fell under the jurisdiction of *Dādgāh-e mobārezeh bā monkarāt* or the Commission Against Social Evils (CASE). When the Ministry of Culture and Islamic Guidance (MCIG) was established three years later in 1982, control over the fledgling video industry was relocated there. This move cemented the state's view of video as closely bound to the film industry. It showed how efforts to control video were part of a larger attempt to regulate formal media systems. During negotiations with the CASE, the MCIG had even claimed that all activities within the field of "film and cinema" were the exclusive purview of the ministry and that the CASE's oversight of video had created uneven cultural policies within the country.[2] Ultimately, it would take the Iranian state nearly five years before it fully regulated videocassettes, first through increased oversight by the MCIG and later through an extensive ban.

Shifting policies, new agencies, and uneven oversight gave the budding video rental market space to grow, and quickly, too. By 1980—just a few years after technology enthusiasts began carrying bulky Betamax play-ers into the country in their suitcases—there were at least ten companies dedicated to transferring and duplicating prerecorded content on video-cassettes. These companies serviced more than one hundred video rental stores, known as video clubs, that now populated neighborhoods in cities around the country. Video clubs cropped up in the most unlikely places: in the spaces under stairwells, in converted garages attached to private homes, and in corner stores.[3] This is consistent with what transpired in other coun-tries, where early video rental initiatives were attached to existing busi-nesses, especially hardware stores.[4]

New trade publications like the magazine *Sinemā dar vide'o* (Cinema on video) and the pamphlet *Vide'o kolub salām* (Video club salaam) provided lists of movies that had received permits for videocassette production. These lists give us a sense of what the video industry looked like in terms of the vid-eos that were legally available for rent and purchase. Of course, these lists do not include the unpermitted movies that likely circulated on videocassette at the time. Nevertheless, they provide a snapshot of the content around which the video industry was developing in the early 1980s. From 1979 to 1983, a mere four years, the government provided distribution permits for approxi-mately five hundred movie titles on videocassette—primarily American, French, Indian, and Italian.[5] There were also some Iranian films with permits for videocassette distribution, including cartoons for children and movies like *Ragbār* (Downpour, dir. Bahram Beyzai, 1972), *Shuresh* (Mutiny, dir. Reza Mirluhi, 1973), *Gavazn-hā* (The deer, dir. Massoud Kimiai, 1974).[6]

To put access to these five hundred titles in perspective, people previously had been limited at any given time to watching just a handful of movies in theaters. Within four years, however, a legal, industrial, and social infrastructure emerged that provided people with a bounty of movie choices. With videocassette distribution centers, a growing number of rental stores, and a long list of approved titles, the video industry was beginning to gain traction in Iranian society. By 1983 videocassettes were still considered a luxury item, but they were becoming commonplace through the formal institutions that allowed for their circulation, including the Association of Video Club Owners and *Cinema on Video*, a publication dedicated to movies on video. No longer confined to society's elite, the consumption of videocassettes even began to permeate the country's middle and working classes. A report prepared by the Association of Video Club Owners, for example, stated that by 1983 a significant portion of video clubs in Tehran were located in the southern part of the city, a traditionally working-class neighborhood.[7] Videocassettes were quickly becoming a widespread phenomenon.

As video gained traction in the country, it also posed a threat to a new government still getting its bearings. Home video technology challenged governmental control by offering citizens the ability to operate outside of formal media systems. During its first decade, the Islamic Republic carefully curated the images and sounds that were allowed to circulate within the country.[8] State-run media, including television and cinema, played an important role in this effort by transmitting images that conformed to the populist, Islamic vision that the state had created for itself. In contrast, the very origins of home video could be traced back to a discourse of consumer democracy that promised to give individuals the freedom to choose how and when they consumed moving images. Early marketers branded video players as time-lapsing devices that freed consumers to watch television programming on their own time, rather than be bound to predetermined timetables. The ability to "Watch Whatever Whenever"—as Sony famously advertised its Betamax player—went against the government's desire to control what people watched and how. Everyday people and small business owners could sidestep the regulatory policies and top-down hierarchies that had largely organized

mass media communication up until that point. At a time when the new government in Iran sought to heighten its regulation practices and capitalize on
the rigid structures of broadcast media, video was an unwelcome addition
to the tightly controlled media ecology that the Islamic Republic sought to
nurture.

If video technology's promise of democracy gave pause to a government
that sought to limit access to media, then the revolutionary rhetoric that
circulated alongside the new medium terrified the Iranian state. Around
the world, popular sources touted home video technology as a revolution.[9]
Such claims could even be found in Iranian trade publications from the
late 1970s.[10] Of course, these sources positioned video as a media revolution that would upend traditional forms of production, distribution, and
exhibition—and not a political revolution like the one that Iran had just
experienced. Nevertheless, the word "revolution" was enough to alarm the
Islamic Republic, which had a contentious relationship with new media. As
Annabelle Sreberny and Ali Mohammadi have shown, homemade audiocassette tapes and xeroxed pamphlets played an important role in the success of the revolution of 1979.[11] As a result, by the early 1980s, the state
understood the power of consumer media technologies like video. Given
the histories of these kinds of technologies in the country, informal media
at this moment held more political potential than simply as leisure, hobby,
or the rejection of capitalistic industry. As one Middle East distributor told
Variety in 1988, "They haven't forgotten the recordings of the Ayatollah's
speeches in Paris before the Shah was overthrown. They don't want anyone
using videocassettes in that way against them."[12] Consumer media technologies had been implicated in a revolution once before. For many within
the government, it wasn't a stretch to imagine that they could be deployed
once again to unravel the new republic's power.

It wasn't just the state's political power that video technology threatened. Support for the ongoing war effort was also in jeopardy. The start
of the Iran-Iraq War in 1980 made it especially important for the state to
broadcast a unified vision of itself. Home video technology, at least in theory, threatened this project by competing with television for consumers.
The distribution of a wide selection of local and foreign movies on video
created more entertainment options at a time when the government was
consciously limiting leisure activities to steer its citizens to consume the
state's wartime ideology on limited TV programming. At this moment, as

the Iranian public grew tired of war, the state's official media were called to arms and asked to keep the ideological project of the war alive. Home video could not be part of that project because its uncontrollability challenged the very principles of state control that drove media production and distribution at the time.[13]

While television in Iran in the 1980s promoted the war effort, cinema at that time sought to attend to the people's spiritual and moral wellbeing. When revolutionary leader Ayatollah Khomeini identified cinema as a tool for educating the masses in 1979, he perhaps unwittingly put forth a new vision for cinema altogether. This imperative placed tremendous pressure on the government to control how people produced and consumed motion pictures in the Islamic Republic. Once the MCIG assumed control over the film industry and after the establishment of the Farabi Cinema Foundation (FCF, *Bonyād-e sinemā'i-ye fārābi*) in 1982, the two institutions worked together to foster an "Islamicate cinema" by capitalizing on the very limitations of the film medium.[14] The FCF controlled access to expensive film equipment and processes (such as developing and editing), while the MCIG controlled what movies people could watch by issuing exhibition permits. Video technology did not require expensive equipment and processes or specialized exhibition venues, so it was a threat to the exclusive movie culture that the state sought to develop—one that existed only on film.

Video not only challenged the state's attempts to control cinema's ideological orientation but also endangered the film industry's financial wellbeing. The violent interruptions to commercial cinema in Iran in the late 1970s and early 1980s—as film transformed from an industry motivated by ticket sales to a state institution at the service of the government's official ideology—presented serious financial challenges. Khomeini's vision for cinema as a propaganda tool presupposed that people would want to watch movies regardless of their content. Reported figures from this period, however, show that box office success was not a given.[15] Starting in the early 1980s, the film industry required heavy government subsidies to survive. The state's new brand of cinema—which consisted mostly of moralistic tales, stories of the downtrodden, and narratives about the perils of corruption—struggled to stay afloat because it could not support itself on ticket sales alone. Iranian cinema of this time became a cinema of "crisis."[16] The drastic drop in movie theater attendance was likely caused by a combination of indifference to the ideologically oriented movies available and the instability of the moviegoing experience, which was interrupted in many cities by bombings and power cuts because of the war.

Home video offered an alternative to these possible deterrents by allowing people to watch the movies that they wanted to at home. The video club industry had begun to form as the sale and rental of videos became profitable. Video clubs unionized, and trade publications emerged to support the technological and economic concerns of the budding industry. Articles like "Filmmaking and the Possibilities of Video" and "The History of Video and Videocassettes" populated these short-lived publications.[17] By 1982, the young industry had begun defending itself against claims that video distribution would threaten cinema in the country. In an article titled "How Serious Is Video's Threat to Cinema?" in *Cinema on Video*, professionals in both the film and video industries speculated on the effect that video would have on cinema. Drawing on examples like the adaptability of Hollywood to new technologies and industries, including television, the article suggested that video actually posed very little threat to the larger project of cinema and might even aid it in its new mission.[18]

Nevertheless, the growing industrialization of video, which included everything from trade unions, organizations, and publications, indicated the two prongs of video technology's economic viability: its success in the marketplace and the growing demand for movies on video. It was precisely the economic desirability of video that troubled the government. The growing popularity of videocassettes could have impeded the development of a new state-sanctioned cinema by diverting ticket sales away from locally produced movies. Video created too much competition in the movie marketplace—competition that the state did not think it could afford at a time when it wanted to encourage viewership of its new cinema.

In many respects, the origins of home video technology in Iran resemble the story of videocassettes elsewhere in the world. A niche market for technology enthusiasts soon gave way to a profitable business venture with specialized stores and growing levels of expertise. Yet as state and society contended with this new medium, its possibilities and its dangers became apparent for both. As is often the case with new media, a period of unregulated experimentation was followed by attempts to rein in amateur practices as governments, industries, and users confronted the ethical, legal, and economic possibilities of the technology. In most places, the stakes of this process for analog video were largely commercial. In the United States, for example, media conglomerates deployed intellectual property rights laws to protect their financial interests. In Iran, however, the informal practices that video technology nurtured did not just threaten profits but also the wellbeing of a young, vulnerable government at war. As a result, there was no place for this unruly medium in the country's formal media industries. A ban seemed the only option.

VIDEO BANNED

The institution of a ban on home video technology between 1983 and 1994 represented the most stringent form of regulation imaginable. It was a bold attempt not just to control a new medium but to deny its very existence. The ways in which the MCIG and other state institutions articulated the video ban revealed the lengths to which the young Islamic Republic went in order to regulate media culture at this time. The state heightened video technology's associations with American popular culture and cultivated a moralistic and anti-imperialist vocabulary in order to dismiss it outright. But harsh rhetoric and stringent laws alone do not make an effective policy. By all measures, the video ban failed. Between 1983 and 1994, video did not just survive. It thrived. The underground video economy that developed at this time was vast and valuable—with an estimated net worth of millions of dollars. In contrast to the state's charged rhetoric, the phenomenal rise of the underground video infrastructure might seem surprising. The state's impassioned critique of video obscured its uncertainties about the policy and the ban's unintended effects. The ban period saw uneven enforcement but also dissenting voices from within the government about whether or not a wholescale ban was the most effective way to manage the unruliness of video. Embedded in the state's motivations and its failures is an important lesson about media regulation in Iran. The government's power is not absolute or unyielding, and even the harshest measures have consequences unforeseen by the state. As I will show, the failure of the ban was the success of video, and the failure of the state a victory of the people.

By the time the MCIG announced the ban, the video industry had been under attack for more than a year. The transition in oversight from CASE to MCIG in 1982 had a devastating effect on video in Iran because the MCIG's harsher standards meant that fewer titles were legally available. As a result, many video clubs were forced to close for financial reasons.[19] In the months leading up to his announcement of the ban, Mohammad Khatami, the head of the MCIG, had disparaged the technology, calling it a disease that could infect all of society. He said, "Currently a plan is being developed to implement precise rules in order to prevent the growth of this issue [video] into a cancerous phenomenon [padideh-ye saretāni], which, without sufficient regulation, could fall into the lap of obscenity and create disasters in our society."[20] Elsewhere Khatami described video as "a harmful entity and the cause of moral corruption and degeneration."[21] By describing video as a cancerous tumor that could metastasize to society's other organs, Khatami's rhetoric in these statements aligned with a longer tradition in Iran of likening media to disease (figure 1.1). The vocabulary of disease, epidemic, and infection had long inflected how Iranian intellectuals and policymakers attacked media, both before and after the revolution.[22] Such rhetoric shows how the cultural values of video in Iran depended on a longer discursive tradition.

When Khatami formally announced the video ban on May 5, 1983, he drew on the rhetoric that had already been used by the state to describe video. He decried video as "a source of moral corruption . . . degradation, and obscenity in our society."[23] The use of morally prescriptive terms like "corruption" (fesād), "obscenity" (ebtezāl), and "degeneration" (tabāhi) had become a common strategy for the new government to dismiss people, entities, and actions of which it otherwise disapproved. Such a strategy drew on the new political authority of religious discourse, specifically Shi'a Islam. During this time, however, there was no official religious jurisprudence on video technology in Iran.[24] Clearly then, this issue was not a point of contestation for the religious scholarly community in the country at the time. Even though the government activated religious rhetoric, the dismissal of video appears to have been primarily politically motivated and not indicative of a religious, scholarly debate.

In addition to religious rhetoric, the state justified the video ban through the language of anti-imperialism. This justification is consistent with revolutionary discourse at the time. The Islamic Republic consolidated its ideology through religious concepts like jihad and martyrdom while also positioning itself against cultural imperialism and gharbzadegi, meaning an obsession with the West.[25] The country's leaders articulated their critique of cultural imperialism through the term tahājom-e farhangi (cultural assault). This was the topic and title of an entire book by Iran's second supreme

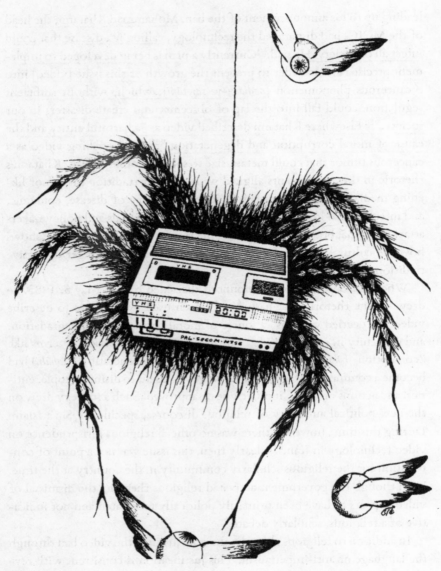

Figure 1.1
Published shortly after the ban, an illustration in the *Sorush*, the official publication of
the Islamic Republic of Iran Broadcasting (IRIB), shows video technology as a predator.
Source: "Tasvir-e ebtezāl mahv mishavad," *Sorush* 194, June 4, 1983, 32.

leader, Ayatollah Ali Khamenei.[26] Cultural assault became a designation for video technology, too.[27]

The state's use of anti-imperialist rhetoric to justify its ban likely resonated with many Iranians. Before the ban, the Association of Video Club Owners undertook projects aimed at improving public perception of videocassettes. This was necessary because many people associated the technology with the United States and Europe and therefore viewed it as being at odds with the values of the revolution, which was anti-imperial in spirit.[28] Following the ban, the state exploited these reservations. For example, the MCIG circulated a booklet in the mid-1980s called *Vide'o va ma'muriyat-e ān dar Iran* (Video and its mission in Iran). It contributed to the discourse on video and imperialism by suggesting that during the revolution the U.S. Embassy had encouraged the spread of video players as a way of distracting people from the revolution. The treatise claimed, "The Great Satan [i.e., the United States] commanded the employees of the American Embassy [*lāneh-ye jāsusi*: the nest of spies] to spread video and to encourage the entrance of video players, their related cassettes, and a variety of movies in order to fill the people's free time with the obscenest movies and to separate them from the Islamic Revolution."[29] While it is difficult to determine whether or not these claims are true, "Video and Its Mission in Iran" performed important work by casting video in the story of the revolution. As a tool of the American enemy, even the luxuries that video afforded were dangerous and contrary to the revolutionary commitment that the state demanded. This document also affirmed the state's fears about video—that it would distract people away from ideologically oriented, formal media systems.

Because home video technology mostly operated in private spaces, the state had to go to great lengths to insist that it was part of the public sphere and, therefore, subject to regulation in the eyes of its citizens. In March 1987, for example, the parliament passed a law regulating clothing vendors, specifically their use of "public space" (*malā'-e 'ām*), outlining punishments for those people who violate "public chastity" (*'effat-e 'omumi*). Unexpectedly, the law included a provision on videocassettes. The provision required that photographs, films, videos, and posters with images that violate the country's modesty laws be destroyed. Regarding video, the law states that "after all objectionable sounds and images have been removed, videocassettes and video devices will be recorded as crimes. The government will then take

possession of them and—with the assistance of the court—transfer possession to educational and scientific centers,"[30] where videocassettes would
be repurposed for educational uses. This provision underscores the state's
attempt to maintain video as part of the public sphere, going so far as to
consider the content of videocassettes as public space and subject to the
same level of regulation as a street vendor, store, or café.

The state's laws, policies, and charged rhetoric suggested a definitive
stance on video at this time. But this coherent message obscured the failures
of the video ban, as well as doubts about the policy within the highest levels
of the country's regulatory bodies. Even within state-sanctioned publications, there is ample evidence that the state failed in its efforts to control
video. The trade publication *Film*, for example, published statistics in 1993
that showed that there was one videotape for every 22.5 people in Iran.[31]
While there were far more videocassettes in the United States (one tape for
every 3.8 people), these numbers still reveal a relatively large number in
Iran, most likely concentrated in urban areas. Similarly, the first issue of the
trade publication *Video-Satellite* in March 1994, published after the legalization of video, claimed that during the ban period the "illegal videocassette
system" had represented a "multibillion-rial [multimillion-dollar] business."[32] This is consistent with what we know about the surge in prices that
occurred after the ban. Shortly after the ban's implementation, the price
of a video player skyrocketed from several hundred dollars to $11,000 (or
approximately $25,000 adjusted for inflation).[33] Such figures are stagger-
ing given inflation and the weakened wartime economy. While these numbers do not tell us much about the nature of underground video culture,
they highlight the ban's failures. With so much money flowing into the
underground video system, the state's attempt to disparage video through
its charged rhetoric clearly did little to disrupt the social or cultural value
of the medium.

The failure of the video ban forced the state to confront what it knew all
along: videocassettes were uncontrollable and their uncontrollability was in
part what made them unique. Video technology too easily slipped through
the cracks and too casually subverted legal, industrial, and individual control. Even before the MCIG officially lifted the ban, voices within the government began expressing their discomfort with the ban, which they saw
as backfiring by forcing individuals and, even worse, families to participate
in the informal distribution of videocassettes. In April 1990, for example,

Naser Akbarzadeh, the head of the Headquarters for the Propagation of Virtue and Prevention of Sin (HPVPS, *Setād-e amr-e beh maʿruf va nahi-ye az monkar*), argued that the complete ban on video had precluded the possibility of controlling a video market in Iran. He proposed an overhaul of the existing policy so that permissible content could begin to compete with the strong underground market.[34] As early as 1992, two years before the MCIG lifted the video ban, the Council of the Islamic Revolution of Iran (CIRI, *Shorā-ye enqelāb-e eslāmi-ye Iran*), the highest legislative body in the country, passed a resolution after "lengthy" and "detailed discussions," encouraging the MCIG to prepare to remove the ban on video, arguing that video could even benefit society.[35] By the early 1990s, organs within the country's regulatory body recognized that a ban may not have been the most effective way to wield control over home video technology.

As the HPVPS and CIRI statements suggest, the government was not in the dark about the failures of the ban to prevent unregulated access to videocassettes. The underground world of video was a public secret that everyone knew but few dared to speak of, especially during the ban period. Nevertheless, the MCIG admitted its knowledge of the underground video infrastructure in the first master set of video policies following legalization, entitled "Policies and Procedures for the Production, Distribution, and Exhibition of Films and Video Programs, and Supervision over the Import of Video Equipment." Published in 1995, the policies acknowledged the rampant growth of video during the ban period. The preamble to this document claimed:

> In our country, for several years now, videocassettes have found their way into people's lives. Because they were not very seriously regulated, quite a number of films successfully appealed to viewers of all ages, who watch the most recent American and sometimes European films. These films find their way to domestic [Iranian] viewers through networks of bootleggers almost immediately after they are screened publicly in the United States and Europe. Acknowledging this phenomenon, it has been decided that video clubs will reopen, and video activities will be legalized through stores that specialize in cultural products.[36]

In just a few sentences, the MCIG conceded the errors of its ways. It acknowledged that a vast informal world of videocassettes had thrived beneath the state's harsh rhetoric. The MCIG's stated motivations for the ban had included protecting the public's moral wellbeing and warding off cultural imperialism. In this brief introduction, the MCIG admitted to the failure of both and acknowledged the need for a new regulation strategy, which would control the distribution of videocassettes rather than attempting to foreclose the possibility of access altogether.

By its own admission, the state did "not very seriously" regulate video during the ban period. This is not to say that the risks of getting caught were not real or that punishments were not severe. Rather, the enactment of the video ban was more uneven in practice than the state's definitive language might have suggested. The MCIG's statement, of course, stopped short of admitting the impossibility of controlling an unruly medium like video through a stringent policy like a ban. As we saw in the previous section, the properties of home video, which put processes like distribution in the hands of the individual, made the technology incompatible with full-scale prohibition. Nevertheless, the state's inconsistent oversight of the ban certainly contributed to the rise of an underground video culture.

In addition to anecdotal evidence, articles published in the American trade publication *Variety* tracked the uneven implementation of the video ban in Iran. In 1984, one article claimed, "Iran's rulers are manifesting a strange tolerance for homevideo. . . . Authorities are closing an eye to this activity as long as homevid peddlers steer clear of Iranian films."[37] Years later, in 1992, an article in *Variety* reported, "Although the authorities frown on the traffic in Western videos, door-to-door rental agents are usually prosecuted only when caught peddling pornography. 'They disappear for a month, and when you see them again their heads are shaved,' notes one local video fan."[38] These reports suggest that while the state may have issued a wholesale ban on video technology, in practice its enforcement often depended on the kind of content being distributed on videocassettes. Another article, published in 1990, quoted a religious official claiming, "banned videos are often confiscated but . . . it's rare for a person to be arrested for watching . . . them in their homes."[39] None of these reports can be taken as hard and fast rules. Instead, they point to inconsistencies in the state's enforcement of the ban. Contrary to the nature of a ban and the MCIG's definitive rhetoric on video technology, the exact parameters of the policy were constantly shifting.

The state may have activated religious, anti-imperialist rhetoric to justify its ban, but that did not necessarily mean that such language convinced everyday Iranians of its importance. The same 1990 article in *Variety* features

Hussain, a "41-year-old business man, who rents several pirated films for his family weekly from one of Teheran's thriving 'underground' video clubs.". According to the article, Hussain "describes himself as a devout Muslim and 'a believer' in many of the cultural and social changes wrought by the late Ayatollah Khomeini." Hussain's religious and political convictions, however, did not necessarily mean that he was willing to abide by all the state's laws, especially those that he deemed arbitrary. He said, "We work hard but they give us so little in return. They say we're not supposed to see this film or listen to this type of music even if we can't see anything wrong with it. I don't care what they say anymore. I'm going to have some fun."[40]

The state's attempts to control video through a ban coincided with larger efforts to regulate media culture at this time. In particular, the government enacted violent measures to reshape cinema according to its new principles. In 1986 the MCIG issued a far-reaching order to collect and destroy unpermitted films. It eventually set thousands of 35mm film reels on fire. Viewing the video ban alongside these film burnings reveals a fuller portrait of the MCIG's regulatory practices at the time.

Ali Abbasi, a well-known producer, recalled the day that the order was executed at his production company:

> The nightmare began on a cold, languid day in February 1985. A colorless pickup truck parked in front of the Payam Cinema Institute. They showed me a piece of paper ordering the search and seizure of all the films in the office. I accompanied them to the first floor, where all of the films, negatives, and ribbons with music and sound effects were carefully stored. As though they had discovered a shipment of cheap knockoffs, they gathered everything and took it with them. Later they would tell me that they set all the films on fire, even the negatives. *Hassan the Bald* [1970], *The Window* [1970], *Reza the Motorcyclist* [1970], *Tangsir* [1973], Abbas Kiarostami's commercials. . . . They took them all.[41]

Abbasi's company, the Payam Cinema Institute (*Sāzmān-e sinamā'i-ye payām*) produced some of the most successful films of the 1970s. The nightmare he described was not just a one-off incident. The order had been handed down by Fakhreddin Anvar, the newly appointed deputy minister for cinematic affairs in the MCIG. That month, government agents took action. They brought 140 truckloads of 35mm film reels to the courtyards of the MCIG and other governmental buildings, including courthouses and the Shah's former palace. After a week-long inspection of the films, law enforcement

officials arrested those individuals who had been storing illegal films. But
the unpermitted reels themselves were left to the elements. Sun and mois-
ture damaged many of them beyond repair. Eventually, government officials set the films on fire, destroying them forever.

It is impossible to know exactly how many films were destroyed in these
burnings. Experts estimate between three and five thousand reels.[42] How
many of those were original film negatives remains unclear. National Film
Archive of Iran (NFAI) officials have claimed that a significant number of
films from the Pahlavi era have been lost forever.[43] While the decision to
destroy unpermitted films has meant a devastating loss for the preservation
community in Iran today, at the time it was a direct assault on the formal
film distribution institutions that had developed over the previous three
decades. Ideological turmoil and the lack of material and financial resources
had meant the decline of Iranian film productions in the first several years
of the 1980s. With few new Iranian productions and no permits for local
films from before the revolution, distribution companies depended on foreign imports. By 1984 more than forty distribution companies dealt in foreign films, many of which had previously specialized in Iranian commercial
cinema.[44]

The newly established MCIG saw these foreign films and those companies
that dealt in them as direct threats to the new kind of cinema it was trying
to develop. As a first step, Anvar required that all production and distribution companies register and document the foreign films they had in stock.
This paperwork laid the groundwork for the confiscation of film reels by
providing a roadmap to privately held archives. From there, it was easy to
seize both foreign films and the unpermitted Iranian films stored alongside
them. Setting the illegal films on fire was a convenience that capitalized on the
highly flammable nature of celluloid film. But it was hardly a coincidence.
The decision to burn the films came on the heels of the revolution, during
which movie theaters were set on fire, most notably Abadan's Rex Cinema
in 1978.[45] The ideological force of the revolution and later new state institutions like the MCIG burned the former film industry to the ground, especially those structures dedicated to distribution and exhibition.

Scholars, commentators, and government officials agree that, for better or worse, these acts of destruction cleared the way for the success of
Iranian cinema in the following decade. Anvar, for example, has defended

his policy, claiming that "if the decision hadn't been implemented, Iranian cinema never would have flourished. . . . In fact, if it hadn't been for this plan, then the base of [today's] Iranian cinema wouldn't have been possible."[46] Scholars, too, have pointed to the fact that cinema and film burnings represented a pattern of "purification" in the years leading up to and immediately following the revolution. In this scheme, the purification of the previous film industry made new industrial and aesthetic practices possible, especially those that furthered the new government's moral and political agenda.[47] In these assessments, the history of Iranian cinema is teleological, as the destruction of one formal industry led to the establishment of another.

It is not that these analyses are wrong so much as they overlook the informal practices developed by ordinary people to access motion pictures. The story of videocassettes in Iran is exceptional not just because of the rapid rates of growth but also because of the incredible circumstances that allowed for them. What is important here is the fact that this media history—the battle between film and home video technology—played out in very material terms in Iran. The material destruction of film distribution and exhibition may have meant the rise of a new, successful formal industry decades later, but more immediately in the 1980s, it created the perfect conditions for a complex network of underground video rental.

Taken together, these extreme regulation measures also teach us something about movie culture more generally during this period. Around the world, cultural critics decried home video technology as the death of cinema.[48] The case of Iran provides an important corrective to the assumption that videocassettes might destroy moviegoing culture. In some respects, the informal world of video saved cinema. At a time when 35mm films literally were being reduced to dust, video as a new medium provided an alternative means of access, preserving movie culture and inspiring new generations of cinephiles.

Movie culture benefited from the informal video infrastructure because throughout the 1980s, cinema as an institution was devastated not just economically but also culturally. During the revolution, cinemas had been caught in the crossfire between revolutionaries, monarchists, and counterrevolutionaries. Movie theaters were vandalized and attacked. Revolutionaries targeted Iranian productions and the foreign imports shown in theaters as hypersexualized and vessels of cultural imperialism. After the revolution, new regulations and the war with Iraq gave cinema little opportunity to reestablish itself. As one cultural critic put it, in the 1980s, once "proud" movie theaters "transformed into nearly abandoned spaces, with posters and pictures that were representative more of an ideological space than anything else."[49] As I have argued elsewhere, it wasn't until after the Iran-Iraq War and Ayatollah Khomeini's death in 1989 that debates within Iranian society and within the film industry facilitated the successful reemergence of cinema.[50] Instead, throughout most of the 1980s, videocassettes incubated a culture of movie watching that otherwise wouldn't have existed.

A critical account of the video infrastructure of the 1980s and 1990s challenges what we know about Iranian media. The scholarship on Iranian cinema, in particular, has presented a teleology of film that valorizes the success of a national cinema. In these accounts, the 1980s and 1990s represented a period of increased regulation. The effort to regulate film-making, however, generated unique industrial practices and aesthetic innovations.[51] This history constructs a narrative about the success of a national cinema, beginning in the late 1990s and early 2000s. This was the kind of cinema that became the darling of the international film festival circuit, represented in pinnacle moments like Abbas Kiarostami winning the Palme d'Or at Cannes in 1997 for *Taste of Cherry* and Asghar Farhadi taking Oscars in 2012 and 2017 for *A Separation* and *The Salesman*, respectively. In the scholarship on Iranian cinema, filmmakers, other members of the film industry, and sometimes even politicians rallied together to create a vibrant cinema, despite stifling governmental oversight. This is an important narrative and it has received a great deal of attention, including in my own previous work. Yet it is hardly complete. The history of videocassettes is Iranian film history's foil. It is the story of the failure of government policy and the failure of movie theater viewership. With these failures, however, came successes. The video infrastructure that developed in the 1980s and 1990s was a complicated and efficient system for media distribution, forged not through state control or incentive but rather through the ambitions, desires, and practices of everyday Iranians at the time.

The video ban bred the perfect conditions for an underground in Iran—especially if we take the underground not simply as a black market for contraband media commodities but as a space where society and state negotiated formality and informality, public and private, and law and ethics. The informality of the video infrastructure depended on the very formal policies of the state, which foreclosed the possibility of a regulated video industry. An absolute policy like a ban created—perhaps counterintuitively—a lot of room for underground practices because such a stringent policy meant that the only way to access videocassettes was through illegal means. These kinds of contradictions extended to the state itself. The MCIG's rhetoric throughout the ban period was consistent, reducing video to a corrupt medium that offended the public's moral sensibility and threatened the country's cultural wellbeing. At the same time, the state's enforcement of the ban proved far less consistent. It often turned a blind eye on the underground dealings that took place. Studies of Iranian media often take the government's word at face value and thus overlook the gray areas that develop between and underneath regulatory policies. The video ban did not merely consist of rhetorical devices aimed at dismissing video technology; it was a set of policies, uneven practices, and rejections from within the government and among citizens.

THE UNDERGROUND RISES

An informal infrastructure for video distribution emerged in Iran not because it was inevitable but because regulatory policies met their nemesis in the videocassette. Unlike today's digital platforms, which governments can filter, or the internet, which they can turn off altogether, analog systems were not so easy to control. The underground infrastructure that developed in Iran was too dispersed and too decentralized to be shut down with a switch. Far from restricting people's access to movies on video, the ban opened an expansive world of informal video distribution that far exceeded the budding video rental industry of the early 1980s. Demand for videocassettes grew rapidly during this period, and a network of underground video clubs emerged to service Iranians' interests in movies on video. Despite the risk of getting caught, video technology flourished throughout the ban period. A robust underground video culture developed as videos were produced and circulated informally outside of state control and formal systems. Instead of preventing the development of an informal world of video, the state had facilitated it.

Statistics about the number of videocassettes in the country and the
financial value of the informal video market gesture toward the expansiveness of home video technology during the ban period. But they do little to
prove just how far the underground extended or to reveal its centrality to
everyday life. In contrast, my corpus of oral history interviews attests to the
pervasiveness of video consumption during the ban period. All the stories
that I heard during oral history interviews underscored the importance of
videocassettes in the 1980s and 1990s. While it is difficult to determine
how widely videocassettes circulated, my research suggests that by the late
1980s, video technology had become a regular feature of life for most mid-
dle- and upper-class urban Iranians.[52]

Given its reach, underground video culture did not just operate at
the periphery of formal media systems. Instead, the informal video infrastructure ran through buildings and homes that outwardly supported the
Islamic Republic and its vision for media. As Ramon Lobato and Julian
Thomas write, "The most important aspect of informal media might not
be . . . [their] exotic, shadowy existence in a parallel world, but . . . [their]
presence—sometimes unnoticed—in the most mainstream parts of the
media industries, and in the most mundane and everyday media experiences."[53] In Iran, the everyday lived experience of video was so mundane
and so widespread that almost everyone seemed to participate, even those
closest to the government.

In April 2013, an Iranian woman named Kowsar Avini shocked readers
by revealing that her father had a large collection of videocassettes when
she was growing up.[54] Given what I have described about the spread of
video technology, this wouldn't be much of a surprise, except that her
father wasn't just anyone. Morteza Avini had embodied the Islamic Republic's vision for cinema in the 1980s. A prolific filmmaker, he directed more
than eighty documentaries about the Iran-Iraq War and blended realist and
mystic-inspired styles that sought to visualize the experience of martyrdom. Such an aesthetic played into the state's religiously charged rhetoric
about the war. Avini was such an important part of the Islamic Republic's cultural landscape that when he died in a landmine explosion in 1993,
Supreme Leader Khamenei declared the day of his death a national holiday,
The Day of Islamic Revolutionary Art (*Ruz-e honar-e enqelāb-e eslāmi*).[55]

That even Avini, who represented a stronghold of state-supported media,
partook in the illegal pleasures of videocassettes shows how widespread

and ordinary the practice was. Although Avini regularly criticized Western culture, his daughter remembered evenings spent with him watching
Hollywood blockbusters like *Terminator 2*. Like most fathers at the time,
he would fast-forward through the inappropriate parts, so that his children wouldn't see them.[56] Kowsar Avini's memories of a childhood filled
with movies suggest how video technology permeated life in Iran, even for
those most closely associated with the government. They also show that
the very formal mechanisms of state-run media—like Avini's documentary
series—were not entirely separate from the underground circulation of
videocassettes.

The relationship between formal and informal media systems ran both
ways. While the case of Avini demonstrates how formal media intersected
with the informal world of video in certain instances, the reverse was
also true. The underground video infrastructure depended on the formal
media industry as well. For example, I interviewed a man named Hushang,
who worked as an employee of the Islamic Republic of Iran Broadcasting
(IRIB). Formerly known as National Iranian Radio and Television (NIRT),
the IRIB is the country's national broadcast agency with exclusive control over the transmission of television and radio airwaves. In the 1980s,
Hushang worked as a television producer, which gave him full access to the
IRIB's resources, including its archives and technical equipment. A self-
proclaimed cinema lover, Hushang began a covert operation converting old
Iranian films, transferring them onto videocassettes, and smuggling them out
of the agency's building. From there, he introduced the videocassettes into
the black market, where teams of video dealers circulated them throughout
the country. When I asked Hushang about his motivations, he told me that
although the work was profitable, his main incentive was to share gems
from the IRIB archives with regular viewers. Given restrictions at the time,
the local film industry was not producing Iranian films on cassette during
this period. Instead, these movies entered the market through intermediaries like Hushang, who used the resources of state-run media to supply the
informal market.[57]

In such moments, to ignore the ban was to reject the subjectivity that
the young government had tried to impose on its citizens. My interlocutors rarely used the word "resistance" to describe their relationship with
videocassettes. More often they preferred the language of escape, pleasure,
and freedom. As one interlocutor told me while holding a small stack of

videocassettes, they represented the ability "to watch whatever I wanted." Nevertheless, the circulation of videocassettes functioned as what James C. Scott calls "hidden transcripts," or invisible, everyday acts of resistance.[58] According to Scott, these kinds of hidden transcripts may not be as public or as seductive for scholars and journalists as a large street protest, but they are more widespread and better indicators of political change.[59] When Iranians threw caution to the wind and ignored the laws forbidding videocassettes, they were not necessarily advocating for regime change. They had, after all, lived through a revolution several years earlier. Instead, they quietly rejected the new social and cultural orders that had come with the political change they had just experienced: an upheaval in class structures, reconfigurations of gender and sexuality, and a growing sense of isolationism.

For those Iranians who felt isolated both from the world and from their national identity in the new Islamic Republic, videocassettes provided an alternative sense of belonging, especially in contrast to the media that propagated the state's ideology. By 1983, radio, television, and cinema were firmly under the control of the IRIB and the MCIG, which had strict procedures in place to ensure that they adhered to the vision that Ayatollah Khomeini and other ruling elite had developed for the country. Other media like satellite dishes and the internet were decades away. Videocassettes, however, were alive and well and offered a welcome escape. The fact that the state disavowed the medium only gave it more credibility. When many urban Iranians brought illegal videocassettes into their homes, they participated in a new underground infrastructure as much as they rejected the infrastructure that carried state-sanctioned signals into their homes.

Through such rejection, media technologies often bore the brunt of people's frustrations. This is consistent with what we know about how people interact with and understand infrastructures. As Susan Leigh Star has shown, a given infrastructure exceeds a single event or site. Infrastructures are "big, layered, and complex" and people "do not necessarily distinguish between . . . [their] coordinated aspects."[60] Unable to grasp an infrastructure's complicated constellation of processes, procedures, relationships, and ideologies, people often register their discontent and their pleasures on its material endpoints. In the case of Iran, people enacted their critiques of the state on media devices themselves. Sometimes that meant simply ignoring the devices. As historian Changiz Pahlevan remembers, in the 1980s "with only a few channels that operated for just a few hours a day, the television was off more often than it was on."[61]

In my interviews, I also heard about many instances in which people attacked their devices as a way of rejecting state media. Kaveh, for example, recalled the day that his grandfather threw their family radio out the window. The grandfather, Ebrahim, was a reserved man who insisted on putting on a suit and tie every morning, even after he had retired from his job as a physician. During the war, an Emerson radio from a decade earlier always played in the background in the living room of their third-floor apartment in northwest Tehran. It had an analog clock, small wooden panels protecting the speaker, and three silver knobs arranged vertically on the right side of the machine. Unlike television, which devoted most of its airtime to news coverage, the radio only featured a few hours of news. This is why Ebrahim preferred the radio to the television and usually spent his evenings in an armchair next to the receiver, where he could listen to classical music. Given his grandfather's special relationship with that radio, Kaveh could never have imagined its fate.

While radio provided a reprieve from the stresses of life in the 1980s, it was also the first means of obtaining information when it came to updates about the war. Although the schedules published in *Sorush*, the weekly (and later biweekly) publication of the IRIB, suggested that news coverage constituted a small portion of programming, in reality news about the war often interrupted musical and other entertainment-oriented programming. Ebrahim often reacted to these interruptions with sighs of disapproval. But one day in the mid-1980s, when a voice interrupted one of Chopin's *Nocturnes* to introduce a live speech by Ayatollah Khomeini, Ebrahim had enough. He calmly stood up, unplugged the radio, and hurled it out the window. He announced that they would no longer be listening to the radio and then sat back in his worn armchair, opened a book, and started to read as though nothing had happened. For Kaveh, who was a child at the time, the whole scene unfolded in slow motion with the *Nocturne* still playing in his head as his grandfather chucked their radio onto the street. Kaveh desperately wanted to look out the window to see the radio in pieces but somehow knew that it was off limits now, even the sight of its destruction. In this instance, the radio set itself was the material embodiment of a state-controlled infrastructure that functioned according more to a logic of interruption than to one of entertainment. Ebrahim destroyed the family radio set to reject the state-controlled programming and interruptions that had come to define it for him.

During my interviews, I often heard stories like Kaveh's. People regularly expressed their frustrations with state-controlled media by assaulting their devices directly. Smacking television sets, ripping newspapers to shreds, and kicking movie theater walls were all narrative tropes that people used to describe their disapproval of state-run media. The media technologies themselves were not only material objects but, for most people, also the only access points to the state media infrastructure, which seemed large and unknowable in its entirety.

This violent rhetoric—as much as it was often tongue-in-cheek—contrasted sharply with the kind of language that my interlocutors used to describe their relationship to videocassettes. When it came to videotapes, they drew on words of love and admiration. They told me stories of diving to catch a falling videocassette, carefully cleaning the head of the tape, and gently blowing into the video player to clear out dust. Where there was anger toward those material media that represented the views and structures of the state, there was also affection for home video technology as an alternative. People bought into the allure of home video technology and participated in the circulation of videocassettes—and thereby established an alternative media infrastructure—in part to rebuff the media infrastructure provided by the state.

In evaluating why videocassettes were an enticing option, we might also consider declining movie theater attendance at the time. Following the revolution, the state poured considerable sums of money into developing a national cinema that aligned with its new priorities. Although the state encouraged its new vision through regulation and subsidies, the results were not always popular. The state could dictate what movies were produced, but it could not demand that people pay to go to the theater to see them. Figures from the time show that theatergoing was not a popular pastime. In the mid-1970s, Tehran housed approximately one hundred and twelve movie theaters, and on average fifty million tickets were sold in movie theaters annually. A decade later, in 1984, there were only seventy-six movie theaters in operation, and only forty-seven million people saw movies in the entire country, with about half of those viewings in Tehran.[62] These numbers show a clear decline in

moviegoing culture in Iran in the late 1970s and early 1980s despite a growing urban population.

As we saw in the previous section, the state may have blamed videocassettes
for the country's ailing cinema. The fact that by 1984 ticket sales had already
been cut in half tells another story entirely. Indeed, even before the large-scale
adoption of home video technology following the ban in 1983, cinema was
already in decline. Home video technology may have contributed to the further decline of the cinema in the 1980s, but people had already abandoned the
movie theater even before videocassettes became a widespread possibility. The
reason for the crisis of cinema likely had to do with a subsequent revolution
and war and a new moral regulation over the film industry.[63] In the 1980s,
as most urban Iranians adopted home video technology, the films released in
movie theaters were poor-quality, didactic films that did not appeal to most
viewers, especially children. This was a sentiment that I heard expressed often
in my interviews. The first feature-length children's movie, *Dozd-e 'arusak-hā*
(The doll thief, dir. Mohammad Reza Honarmand), was not released in theaters until 1990. Up until that time, most of the Iranian films produced were
supposed to be suitable for the entire family. Although they often featured
children in the main role, they were actually bleak tales of the downtrodden,
and few children had the patience to sit through them. As one of my interlocutors Payam declared, "We didn't go to the movie theater, because the movies
there were shit [*mozakhraf*]."

Payam's father, however, had a slightly different take on cinema's declining popularity. For him, as much as the movies in theaters may have been
mediocre, the real reason for staying at home had to do with the war happening around them. With bombings and missile attacks possible at any
moment, traveling to movie theaters with the entire family seemed like
an unnecessary risk. Even if people could get to the theater, the experience was often unreliable, since the power could cut at any time. On top of
these safety concerns, money also played a role. With an ailing film industry and a wartime economy, ticket prices rose consistently throughout the
1980s and 1990s,[64] and it was still relatively expensive to go to the movies. The price of a single ticket in the mid-1980s was 500 tomans, or the
same price of renting five videotapes that the entire family could enjoy. As
much as videocassettes represented an alternative to other state-controlled
media, they were also a technology of convenience. With a well-calibrated

delivery system, a wide and constantly evolving selection of rental choices, and a reasonable price, home video technology was an option that many Iranians were willing to buy into, even if it meant breaking the law.

In Iran, watching movies on video became just as frenzied a pastime as watching soccer. The entanglement of these two pastimes was seen on an episode of *Navad* (Ninety), a TV talk show dedicated to soccer and arguably one of Iran's most popular on-air programs. On November 6, 2017, soccer player Omid Singh joined host Adel Ferdosipour (figure 1.2). Singh, who was born in Iran to an Indian father and an Iranian mother, inspired a number of questions from Ferdosipour about his heritage. The discussion of India and Singh's last name also prompted Ferdosipour to share a memory of videocassettes. He confessed that in middle or high school, he would watch and rewatch the Indian crime thriller *Sholay* (dir. Ramesh Sippy, 1975) on videocassette. With the kind of enthusiasm that Ferdosipour usually reserved for describing the Iranian national team's victory in some international match, he narrated how he had watched the movie so many times that he had it memorized "from start to finish." At a certain

Figure 1.2
During an episode of *Ninety*, Adel Ferdosipour and Omid Singh take a break from discussing soccer in order to chat about *Sholay* on video.
Source: "From Gabbar Singh to Omid Singh," *Navad*, November 16, 2017.

point, though, Ferdosipour became acutely aware of what he was doing: using Iran's most public forum to discuss a very private, illicit experience. The ban on videocassettes had been lifted more than two decades prior. Yet from the perspective of state-run TV, there was still something improper about discussing the pleasures of his videotape crimes on air.

This awareness plays out in the episode as Ferdosipour describes the scene that moved him most. At the end of *Sholay*, one of the main characters Jai (played by star Amitabh Bachchan) meets his untimely death. Ferdosipour explains that at the moment of Jai's death he "would always do *this* to the video." He never clarifies what he means by "this." Instead, caught describing the mechanical pleasures of videocassette technology, he pauses and diverts course. He continues, "it's so sad. Every time Jai died, I would cry. When Jai got shot, the video was fuzzy (*barfak shod*)." We never learn what Ferdosipour would do to the videocassette at the moment of Jai's death. Would he pause the video? Fast-forward it? Or stop it altogether? The particularities of videocassettes as a home viewing technology get obscured in his emotions, inextricable from the tears he would cry every time he watched the movie. Here the magnetic tape's breakdown—the *barfak*—was an extension of his crying. His reaction to the film—whether it was fast-forwarding or pausing on this scene—conditioned a particular response from the technology. Breaks and visual disruptions like these were not distractions from the viewing experience, but rather were embedded into the experience, part of videotape pleasure itself.

What we learn from Ferdosipour's on-air confession is how important videocassettes became to people's lives, especially in the 1980s and 1990s. Ferdosipour was born in 1974, so the period he describes was likely the late 1980s or early 1990s. During my interviews, almost everyone from Ferdosipour's generation remembered watching *Sholay* on video. Videocassette technology may have been a way to consume Hollywood movies. But, as became apparent in Ferdosipour's commentary, it was *Sholay*—an Indian action thriller from the 1970s—that came to define home video for a generation of Iranians who grew up during the Iran-Iraq War. As I showed, the government claimed to ban home video technology in an attempt to safeguard against cultural imperialism, especially from the United States and Europe. The fact that a movie like *Sholay* would find popularity alongside contemporary Hollywood blockbusters shows how videocassettes were always a global phenomenon. Although the state was worried about videocassettes as vehicles for cultural imperialism, people's actual viewing practices told a different story. In my interviews, videocassettes never represented the interests or values of a single nation. Instead, they were, as a whole, the material manifestation of the world itself, a way to access different corners of the globe from Iran at a time when the feeling of isolation ran high.

VIDEO LIVES ON

Although the ban itself lasted only a decade, the underground infrastructure that it spurred would last for many more. In fact, it persists today, determining how people access and consume media and functioning as a site upon which people continue to negotiate informal and formal media practices, just as they did during the ban. The failure of the ban to regulate video technology throughout the 1980s and early 1990s forced the state to recognize what it already knew: that video was uncontrollable but could be harnessed for a profit. The video economy throughout the ban period had proven the financial viability of the medium. With this in mind, the state shifted its approach to regulating video beginning in 1994. Rather than try to outlaw an entire medium, the MCIG focused its efforts on limiting the content that was formally circulated through that medium.

The legalization of videocassettes and players, therefore, did not mean an unregulated formal video industry. On the contrary, the MCIG founded a video distribution company that would represent the state's interests and hold a monopoly on formal videocassette distribution for many years to follow. Yet just as the video ban had done little to prevent the informal spread of videocassettes, so too did the state's monopoly on formal video distribution fail to disrupt the powerful mechanisms of the underground video infrastructure. Given the new restriction on videocassettes, the informal infrastructure continued to play an important role in how people circulated and accessed movies on video. By 1994 the underground video distribution network was vast and efficient. Despite the government's best efforts, a sudden shift in policy could not shake it.

Shifting cultural policies were not uncommon at the time. The changing of the guard in the MCIG—following the resignation of Mohammad Khatami in 1992 and the appointment of a new minister, Ali Larijani—inevitably led to reconfigurations of old policies. The early 1990s also marked a unique period in Iranian history. The end of the Iran-Iraq War in 1988 created a demand for more personal freedoms, and the death of Ayatollah Khomeini in 1989 gave rise to factionalism in the country, as competing political groups sought to make sense of and capitalize on Khomeini's masterplan for the republic in his absence. The late 1980s and early 1990s, therefore, witnessed several changes to the policies that regulated cultural life in the Islamic Republic. Scholars and critics regard this period as a "golden era" for cultural production in the country, as the MCIG relaxed its regulatory practices and permitted the distribution of a number of controversial works, including director Mohsen Makhmalbaf's film Nowbat-e 'āsheqi (A time for love, 1989) and Shahrnush Parispur's novella Zanān bedun-e mardān (Women without men, 1990).[65]

While the transformations to the state's stance on video were part of a larger pattern of cultural policy change, they were still motivated by reasons specific to the discourse on video technology. The state's desire to legalize video was an effort to participate in the profits of the underground video economy. The need to rebuild the postwar economy during this period was especially urgent and in line with the liberal economic policy that the president, Akbar Hashemi Rafsanjani, sought to enact at the time.[66] The state's economic priorities were revealed in its emphasis on controlling only those videos destined for the marketplace. During the decade-long ban, the discourse on video technology changed very little: it was a tool of cultural assault that could be easily misused. By the mid-1990s, however, Iran's damaged economy demanded that the state abandon its religious, anti-imperial rhetoric to pursue economic stimulus through video. The "Policies and Procedures for the Production, Distribution, and Exhibition of Films and Video Programs" even claims that one of the main advantages of video technology is the fact that capital investment can be recovered and multiplied much more quickly than television and cinema. Video was, therefore, "from the perspective of commerce, better" than television and cinema.[67] At the time of the legalization of video, certain editorialists expressed concern over how much money the government was preparing to invest in videocassette technology as it established formal procedures for the production and sale of videocassettes.[68] Yet the state remained resolute in its decision precisely because of the financial success of the underground video system.

The economic viability of video was such a priority for the MCIG that one of its first initiatives in preparation for the legalization of video was the establishment of the Visual Media Institute (VMI, *Mo'asseseh-ye rasāneh-hā-ye tasviri*), a semi-private organ of the ministry. The VMI's main imperative is to provide "financial assistance to cinematic and audio-visual activities," and between 1994 and 2000 it was the only video distribution company in the country, overseeing the legal distribution of all movies on video, including domestic productions and imports.[69] The VMI's establishment marked the legal institutionalization of video in Iran, and other initiatives soon followed to promote the video production that the state hoped to facilitate. The Organization for Islamic Propaganda, for example, announced the first Sureh Video Festival, which encouraged amateur video productions (including fictional narratives, documentary, and animation) in three categories (short, medium, and long).[70] The state also offered subsidies to support straight-to-video movies.[71] Whereas previously all video-related activity had constituted an informal system because it functioned outside of state and industry regulation, the state's new approach to video created new tiers by fostering spaces for video to operate legally under the conditions of state control.

Perhaps unsurprisingly, the state's attempts to institutionalize and incentivize video production and distribution did little to break up the underground video infrastructure in Iran. As Star has observed, infrastructures, once in place, are "never changed from above."[72] Rather, the underground video infrastructure adapted and continued to operate alongside the formal industry that developed for home videos. The MCIG's stringent regulations for issuing permits meant that the demand for movies on video still far exceeded the supply available. Video dealers continued to operate, activating the same processes and networks that had developed during the ban period, even as the technology changed from analog to digital and the format from videocassette to video compact disc (VCD) and digital video disc (DVD). The VMI's close affiliation with the MCIG meant that the sanctioned supply of movies on video was heavily censored, with only certain permitted films available for purchase or rent. Most of these were either recent Iranian productions or censored foreign films. The informal video infrastructure continued to operate in full force, providing viewers with movies that they would never find on the VMI's official lists, including uncensored, recent Hollywood hits and old Iranian movies.

Even in late 1993, as the government was preparing to lift the video ban, people—especially those closest to the film industry—understood that the legalization of videocassettes would not necessarily spell the end of underground video distribution. For example, one author writing in *Sorush* contemplated a list of sanctioned films that might both encourage the film industry's growth while also hindering the underground video market. At that point, no one knew which films the MCIG would authorize for video distribution. The author lamented, however, that it was common knowledge at the time that "after legal video operations begin, the unpermitted distribution [of videos] will remain active." "Why," the author asked, "won't any of the entertaining movies that people seek out on video be available in the official video market?" At the heart of this debate was the tension between "the ability to choose" (*now'-e entekhāb*) and a "cultural assault from the West" (*tahājom-e farhangi-ye gharb*).[73] A video market prioritized the ability of consumers to choose what they watched. This was inherently at odds with the state's cultural policies, which aimed to protect citizens from certain media content.

Indeed, discussions about the legalization of video often took shape around the "ability to choose" as an ideal that the formal industry would fail to meet. For example, in late 1993 *Sorush* solicited opinions from Iranians about the upcoming legalization of video. One forty-five-year-old man wrote, "If you want the truth, I have no idea what they mean by 'legalizing video activities.'" His confusion grew out of an uncertainty about what "legal" would mean in practice, especially in terms of the selection of films available. "I remember," he wrote, "the video clubs that operated freely immediately after the revolution. They had every imaginable film available. . . . The most pressing question for me is what kinds of films these new video centers will feature. As long as that remains unclear, I can't offer a definitive opinion on the legalization of video."[74] Because *Sorush* is the publication of the IRIB, there were limits on what it could publish. This man does not reference the underground video market but rather the video clubs that operated freely before the ban. Nevertheless, his message is clear: the new video industry is likely to be limited, especially compared to the video clubs before the revolution and the expansive underground video market operating at the time.

Other participants in the *Sorush* survey found different strategies to express their concerns about the ability to choose, without directly implicating the underground video infrastructure. A seventeen-year-old student, for example, wrote, "The legalization of video clubs is a good step. We have a video player at home and really want to watch movies on video that are not broadcast on TV or shown in movie theaters. I hope that video clubs rent out the kinds of films that young people like me enjoy. I'm particularly interested in detective films, and I can't imagine that there would be any hurdles in distributing them on video."[75] Like almost all of the participants in this survey, she expresses concern about exactly which movies will circulate through this formal market. She implicates the underground video infrastructure when she admits to owning a video player and to her fondness for detective films. In almost all of the opinions in the *Sorush* survey, people grapple with the new formal video industry implicitly through their experiences with the underground video infrastructure that operated robustly by that time.

Before a formal system of video distribution was set in motion—indeed even before a list of permitted movies had been released—trade publications like *Sorush* resigned themselves to the reality that the informal and formal video industries would operate in tandem. As one author wrote, "if video distributors are not able to produce their fine films through legal video centers, then they will gravitate towards illegal means."[76] According-ing to these accounts, the logic of the market would ultimately win. Both consumers and distributors would seek out either a formal or informal video market, depending on their priorities. The formal market had the benefit of being legal, but it was also limited. Meanwhile, the informal infrastructure that existed by 1994 was convenient and expansive, if not entirely legal. It is possible that this dual system would have still emerged if the ban on video had never been implemented. The reality, however, is that by the time the state began setting up a formal video industry through institutions like the VMI, a well-calibrated, efficient underground video infrastructure already existed. It had been providing millions of Iranians with access to movies on video for over a decade. Under those conditions, it is impossible to imagine that a new formal industry could exist entirely separately from the informal infrastructure that had developed to equip people with videocassettes.

The dominance of the informal infrastructure emerged quickly during the course of my oral history interviews. When I asked my interlocutors about the ban period, they would often contend that there had never been a complete ban on video technology. These were the same people who had participated in the informal distribution of movies on video, either as deal-ers or consumers. For example, one video dealer, Hussein, told me that the formal video industry had been closely regulated and therefore limited throughout the 1980s, but that the state had never banned video technology outright. While it would be easy to brush aside these statements as misin-formation, they reveal something much more profound. Many Iranians—especially those who were not attuned to the constant shifts in the state's cultural policies—did not notice the emergence of a new, state-sanctioned video industry beginning in 1994. On a practical level, most people contin-ued to depend on the underground video distribution network in order to exercise their ability to choose movies on video.

People nevertheless visited the official video clubs that began cropping up in 1994. In fact, some of my interlocutors very proudly showed me their VMI membership cards, which allowed them to rent movies from the offi-cial clubs (figure 1.3). Because brick-and-mortar stores had been so rare dur-ing the ban period, the formal video clubs of the mid-1990s were a novelty. People enjoyed browsing for movies—an experience that seemed luxurious compared to the hushed and hurried process of selecting videocassettes from their video dealer's briefcase. At the same time, they understood that sanc-tioned stores served a very different function than the underground video market. Video clubs became a place to browse movies, to perform movie culture publicly, and to procure high-quality copies of a limited number of titles. The underground distribution network, however, fulfilled people's desires; it was a system of convenience, of private consumption, and of a world of movies disentangled from the state's regulatory processes.

Over time, the legal video stores of the mid-1990s and early 2000s became nodal points in the video infrastructure that operated underground. While the first generation of video stores were run by Iran-Iraq War vet-erans close to the government, with time the MCIG loosened its control over the video industry and anyone with the means could open a video club aimed at selling and renting videocassettes. With less direct oversight from the MCIG, some video stores became places where one could find permitted movies on display and a stash of uncensored and unpermitted films tucked away under the counter. As Homayun, a self-described cine-phile from Tehran told me:

> In 1996, when I was 20 years old, a new video store opened in my neighbor-hood. At first, when I would go, the owner would just offer me the standard

Figure 1.3
In *VHS Diaries*, documentarian Niyaz Saghari shows several membership cards for
cultural organizations, including VMI.
Source: VHS Diaries (dir. Niyaz Saghari, 2018). Courtesy of Niyaz Saghari.

stock of videocassettes that you could find anywhere in the city. After several
months of visiting the store and talking about movies with the owner, he let
me in on a secret. He had an entire stock of hard-to-find movies behind the
counter. They were, of course, much more expensive than the videocassettes on
display. He told me that if I ever had a specific movie in mind, to let him know
and he'd try to get it for me.

This story was in line with others that I heard in which video store owners became intermediaries between video dealers—and by extension the
underground video infrastructure—and consumers. Their public-facing
stores were the perfect fronts for the intersection of formal and informal
industries, operating side by side and hand in hand.

Two factors motivated video store owners to deal in illegal movies
alongside their permitted stock. First, the formal video industry struggled
financially. Although the first generation of video stores were subsidized
by the state, later iterations were subject to the dictates of the market.
Because the ban period had trained millions of video consumers to access

cassettes through informal means, they were reluctant to partake in the formal industry that the state abruptly sanctioned. Video store owners knew that they had to offer something extra in order to remain competitive. Second, the video store owners I interviewed all reiterated that they had gotten into the video business because they loved movies. They felt restricted by the limitations imposed by the government and wanted to offer what they deemed to be an appropriate selection of movies on video. It is important to bear in mind, however, that the underground portion of their businesses constituted a very small part of their dealings in video. The formal industry reigned supreme in these shops. As Homayun's story shows, video store owners only offered their selection of underground videocassettes to customers with whom they had a well-developed rapport. Ultimately, the fact that even legal stores became implicated in the underground video infrastructure exposes the slippery boundaries between the informal and formal practices.

As I've shown, the video ban catalyzed an underground where informal media prevailed but never separately from formal policies, institutions, and practices. While the underground may have been born of videocassette technology, it still exists today as a space where people access and consume media by negotiating formality and informality, even as analog video technology has given way to today's digital devices. In the summer of 2016, I settled in with a friend at his home in Yazd to watch *50 kilu ālbālu* (50 kilos of sour cherries, dir. Mani Haghighi, 2016). I had seen the film months earlier during a trip to Tehran. My friend, however, had missed his chance. Just a week after I saw the movie in the Kourosh Cineplex in northwest Tehran, officials plucked it from theaters, bending to pressure from critics who objected to its comedic representation of both marriage and the government's laws regarding contact between men and women.[77] Nevertheless, my friend easily secured a copy from a movie expert who dealt in flash drives and DVDs.

My friend opened the movie's file on his laptop, which was connected to a TV set. He pressed play, and a series of abrupt movements and hushed sounds gave way to a crowded movie theater with the screen at the center. As the lights dimmed in the theater, the camera zoomed in on the screen. Likely shot on a smart phone ill-equipped to film in the dark, the opening credits were cloaked in a soft, unfocused blur. I sighed, ready to give up. I figured the copy was a dud. My friend retorted, "It doesn't get better than this! It's like we're right there in the theater with them." We watched the entire movie, even as it was constantly interrupted: as people stood up to exit the theater, as the image moved in and out of focus, as the audience members' laughter and chatter drowned out the film's sound, and as the person recording it regularly lowered the camera down to avoid getting caught.

What made my experience watching *50 Kilos of Sour Cherries* in Yazd particularly poignant was the fact that the film features an underground videocassette—a scene that I discuss in more detail in chapter 5. When my friend saw the videocassette in *50 Kilos of Sour Cherries*, he couldn't resist drawing a connection between the video culture of the 1980s and 1990s and our experience watching the movie that day. As the characters on screen carefully placed the cassette into an old video player, my friend elaborated. "You see," he explained, "we Iranians are used to watching movies like this. Over the years, we've gotten used to it. Sometimes movies aren't watchable, but one has to watch them anyway." Statements like this bridge the video infrastructure of the late twentieth century with contemporary distribution practices. While it might be easy to view digital and streaming technologies as important ruptures, history tells us that new media technologies often provide more in terms of continuity than in change. Certainly, this is the case in Iran, where the nuts and bolts of the underground infrastructure continue to operate.

An official version of *50 Kilos of Sour Cherries* would eventually be released both on DVD and on streaming sites—this time with its objectionable scenes removed to appease the religious authorities who had opposed the movie. Meanwhile, the movie theater copy that I saw in Yazd continues to circulate widely online under the title *50 kilu ālbālu bedun-e sānsur* (50 kilos of sour cherries uncensored). The multiple versions of the film instruct us on the coexistence of informal and formal modes of distribution, which exist alongside one another, each with its own specific set of affordances. Official releases might promise high-quality images and sounds, but informal versions allow for cheap and uncensored copies to circulate. Just as the ban caused a surge in the demand for videocassettes, the state's formal attempts to regulate *50 Kilos of Sour Cherries* by removing it from theaters actually created a market for uncensored versions to circulate informally.

It would be easy to say that much has changed in the three decades since Nahid and Behrooz secretly traded movies on video in their government offices. Bulky videocassettes have given way to weightless files, and the ban that drove their exchange underground no longer exists. Yet in some respects, very little has changed in terms of how people consume media. Although the underground was forged of hard plastic bricks, it continues to structure media access as a place where formal and informal practices shape one another—and get tangled up like a cassette's magnetic ribbon in an old video player.

UNDERGROUND NETWORK: COLLECTIVITY AND THE VIDEOCASSETTE INFRASTRUCTURE

Simin will never forget the day she got arrested. A dutiful government employee who had a sixth sense when it came to discerning the state's laws at any given moment, she had always been a stickler for the rules. The last thing she expected was to find herself in an interrogation room on a cold day in January 1987, especially while her husband and young daughter waited for her, unaware that she had been dragged to a police station far away from home. And yet there Simin was, begging and pleading with the morality police, trying her best to convince them that she had only been at the video store to visit a friend.

The evening had started off innocently enough. When her husband, Ahmad, came home from work, Simin decided to take advantage of a recent ceasefire to visit a friend who owned a small video store in northern Tehran. At various points during the war, Iraq and Iran bombarded each other's major cities for weeks, sometimes months, at a time. These attacks were called the War of Cities (*jang-e shahr-hā*), and they were brutal. Relief only came when the militaries had temporarily exhausted their resources. Simin was always amazed at how quickly people recovered from those periods of intense violence. Only two days after she, Ahmad, and their daughter, Laleh, had huddled together in their building's basement, certain that night was their last, Simin carefully applied just enough makeup not to get caught as she made the short journey north.

It wasn't just that Simin needed to get out of the house. After weeks of worrying about life and death in the most visceral way, she also craved art and culture. She needed an escape from it all—from piecing together dinner with whatever rationed food was available, from praying that a warning siren wouldn't go off as soon as Laleh had fallen asleep at night, and from worrying about Ahmad every time he was late coming home. She wanted to retreat deep into her mind, someplace where the exhaustion of her day-to-day life couldn't reach.

Luckily, her friend Amin shared her interests in the arts, especially cinema. He was also a former flame—although that was far in the past. Amin had funneled his interests in film into a small, exclusive video store carefully tucked away atop a modest four-story building in the northern stretches of the city. Unlike the video dealers who normally distributed videotapes, Amin's video club was an actual store where members could come and browse high-quality cassettes. Membership was limited to friends and friends-of-friends: the kind of people you knew you could trust.

When Simin stepped into the store, which was really no bigger than a small bedroom, she felt at home among the carefully displayed videocassettes. Amin was busy helping a customer look for something for his niece and nephew, who would be visiting later that week. Simin struck up a conversation with the only other customer in the shop. It was the kind of place where you could talk to the people browsing, and that's why Simin liked it so much. The customer was there for the first time, and Simin explained that everyone had to pay a membership fee, just to make sure that the videocassettes were returned. The woman smiled and nodded. Simin could tell she understood how valuable the cassettes were and the reason for the extra fees. Browsing cassettes like this in a store was a luxury to which few had access.

As Simin and the woman settled into a lively conversation about their favorite movies, the door to Amin's store swung open. Simin casually turned her head to the right to see who was there, and she froze as four or five members of the morality police (*komiteh*) poured into the tight space. Several of the officers began collecting the videocassettes and throwing them violently into plastic garbage bags. Two female officers came up to Simin and the other customer, demanding to know what they were doing in such an obscene store. The other customer managed to convince the officers that this was her first time there, and they let her go. But Amin and Simin were loaded into separate cars and brought to a station a long distance away. After several hours of interrogation, the officers finally accepted that Simin wasn't directly involved in distributing the videocassettes, and they let her go, too. It was already one o'clock in the morning, and she had a lot of explaining to do when she got home.

Amin fared almost as well. His sister was a lawyer and managed to get him dismissed with little more than a slap on the wrist. Unlike other people in his position, he was spared lashings and jail time, but the police confiscated his entire stock of videotapes. Amin never opened the door to his shop again. Instead, Simin helped his wife pack up what was left of the store. As they departed one last time, Simin felt a pang of sadness. She was losing her refuge, her escape from the emotional demands of everyday life in wartime. It wasn't so much that she or any of Amin's other customers would be deprived of movies on video. Rather, it was the fact that brick-and-mortar video stores were unheard of. Amin's store had been a place to gather, to talk movies, and to forget the war outside—if even just for an evening.

Simin was certain that the cassettes the police had confiscated from Amin's shop would end up on the black market and back in the rental circuits that sustained so many people in those years. In fact, every time her family's own video dealer showed up, she carefully inspected his collection, looking for the familiar mark of Amin's handwriting on the labels. She felt that there had to be a force to the underground video network that would deliver one of the cassettes from her favorite hangout back into her hands. Simin never did find a cassette with Amin's childish handwriting on the front, but to this day she still isn't convinced that one of his cassettes didn't end up finding its way home. They had all felt so connected through video-cassettes in those days. Plus, her one and only arrest story needed that kind of poetic justice as its ending.

By the time Simin was sitting in that cold, dark interrogation room in January 1987, the underground video network was already vast. And it was growing larger by the day. As I showed in chapter 1, the laws and policies banning home video technology represented the state's official stance on media but did not necessarily curtail or even limit people's access to movies on cassette. Simin's experiences—like those of almost everyone with whom I spoke—showed that a lively video culture existed in Iran in the late 1980s, complete with multiple access points, modes of engagement, and consumer buy-in.

In the span of four short years, between the banning of video technology in 1983 and Simin's interrogation in 1987, a lot had changed. Video culture had transformed from a burgeoning formal industry into a vibrant underground practice. Between 1979 and 1983, specialty video shops cropped up in cities around the country, especially in affluent neighborhoods. As video stores became more commonplace, attempts to regulate the industry and formalize it through unions and trade publications became more widespread. The video ban redirected this energy underground. The video culture that developed there from the mid-1980s through the late 1990s was marked by informality, increased accessibility, diffuse distribution methods, and a growing sense of connectivity.

While the state may have categorized this underground world of videocassettes as chaotic and criminal, Simin's experiences demonstrate that it was highly structured, created through ritualized practices and processes. Her conceptualization of the underground system as a network, one in which people felt connected, however abstractly, through their engagement with videocassettes, invites us to think more about the infrastructure that undergirded this video culture in the 1980s and 1990s. The most pressing questions here are why and how this underground infrastructure for video rental developed. What kind of technological affordances shaped it? What did the underground infrastructure look like? And how was it responsive to and representative of the political, social, and material world that it occupied?

What is at stake in this chapter are the conditions, processes, and labor that facilitated the movement of sounds and images on videocassette in the 1980s and 1990s. Simin may have felt a sense of connectivity through her access to and consumption of movies on video. But that feeling—which I encountered in many of my interviews—became material through actual technologies, routes, and procedures that allowed for the movement of movies on video. The system of underground distribution structured people's encounters with media and with one another.

Like its English counterpart, the Persian word for infrastructure, *zirsākht*, points us to an underworld; the word "under" (*zir*) is built directly into it. From undersea cables to antennas dressed up as trees, media infrastructures often take shape through a "politics of infrastructural visibility."[1] Such politics reveal how the decision to hide infrastructure—to drive it underground or out of sight—often takes place through negotiations between a variety of public and private entities, including governments, industries, corporations, and local activists and communities. Iran's video infrastructure grew underground because of the same intersection of public and private interests. On the one hand, formal state policies banned videocassette technology; on the other hand, informal entrepreneurs and private citizens capitalized on the particular affordances of home video technology to forge a new system of media distribution. Iran's video infrastructure was underground not because it was invisible but rather because its visibility was contested and negotiated through state policy and everyday practice.

In this chapter, I show how the circulation of videocassettes and the material traces left by that circulation created the idea of an underground network that was simultaneously collective and diffuse. The oral history interviews that form the basis of this book brought me not to a sense of unified national resistance—as I had expected—but rather to the fringes of society: to dark basements, to the porous border between Iran and Pakistan, to the shores of the Persian Gulf and the warfront on the border with Iraq, and to ethnoreligious minority communities. As much as videocassettes pervaded the lives of most Iranians, their circulation originated and proliferated far from the center. At a time when the recently formed Islamic Republic was trying to shape a new national identity by controlling media, the story of video is about ordinary people who sought entertainment and escape. In the process, they forged networks, communities, and practices, which existed outside of the new subjectivity that the state had imagined for its citizens.

IRAN'S INFRASTRUCTURAL INVISIBILITY

Beginning in 1983, as the burgeoning video industry collapsed, amateurs, entrepreneurs, and ordinary citizens configured an informal, makeshift infrastructure to ensure continued access to videocassettes in the country. An illegal but complex underground system for video distribution emerged. The ban may have driven video underground but it thrived because of home video technology's specific affordances. Videocassettes and recorders were consumer technologies that did not require the machinery or resources of a large institution for the production and distribution of prerecorded content. The ability to record and rerecord meant that a robust rental network developed as video dealers made the most of a limited supply of equipment while maximizing both access for their customers and profits for themselves. As a result of both the legal proscriptions of the time and the material properties of home video technology, the underground network was just as expansive as it was fragmented.

The videocassette infrastructure operated according to a logic of invisibility. It aimed to keep videocassettes hidden but also remained hidden and decentralized itself. Diffuse and vast, the underground infrastructure safeguarded against the state's careful watch, while also reaching wide and far. The underground network in Iran therefore provides us with an opportunity to revisit one of the central tropes of infrastructure theory. Invisibility is an organizing principle in many of the theoretical engagements with infrastructure.[2] This is a genealogy that takes seriously Susan Leigh Star's claim that infrastructures are, by their very nature, invisible and are "rendered visible" only "on breakdown."[3] As a result, one of the main intellectual projects of infrastructure studies has been to excavate the complex processes and materials that organize everyday life. Such studies make visible the structures of power that infrastructures work to naturalize. This line of scholarship extends Bruno Latour's claim that infrastructure is a "black box" that needs to be made legible.[4] Thus, much of the field assumes a rigid dichotomy between invisibility and visibility.

At the same time, the ethos of invisibility that undergirds much of infrastructure theory overlooks "the range of visibilities that move from unseen to grand spectacles and everything in between."[5] Increasingly, scholars are showing that invisibility is not a default condition for infrastructures, especially outside of the United States and western Europe. From exposed wires to jailbreaking software, ordinary users across the Global South see and contend with infrastructures on a regular basis.[6] The underground distribution network in Iran was one such infrastructure. It operated within and between different kinds of visibilities and invisibilities. It was the embodiment of what Lisa Parks calls "the politics of infrastructural visibility," as its visibility was contested, negotiated, and manipulated for its many stakeholders.[7]

To understand the underground network against infrastructure theory's invisibility metaphor involves examining how and why it was made invisible and for whom. To survive, the underground video infrastructure in Iran needed to remain diffuse and obscured for the people who accessed it. A clear set of procedures and protocols would have made it more vulnerable to punishment and state control. However, the diffuse nature of the underground network did not mean that users were unaware of it or that they did not see parts of it. Infrastructures are too vast and too complex to be visualized in their entirety. But the idea of media infrastructure often exists vividly in people's imaginations, especially in contexts like Iran in the 1980s and 1990s, where underground video consumers had to confront the material and legal dimensions of media access on a regular basis.[8]

Thus, like all media infrastructures, the mechanics and technical workings of the videocassette underground were pushed far out of sight for most users. This kept people in the dark about how it operated but also enticed their imaginations, which became nearly as important as the reality itself. With limited sources, mostly oral histories, my own attempt to render this infrastructure visible is necessarily incomplete—and perhaps at times even incorrect. But this is how it existed in people's imaginations; how they understood this material world during a time of tremendous social and cultural change.

Although the underground infrastructure was necessarily invisible due to its illegality, it was in some ways a continuation of the burgeoning legal video industry of the early 1980s. Infrastructures are large, complex systems. They do not just appear out of thin air. Brian Larkin, for example, proves the continuity between various infrastructures in Nigeria over time, as nineteenth-century colonial projects like railroads and highways ultimately cleared the way for an indigenous movie industry in the 1990s.[9] In Iran, a similar continuity occurred, even though many of the people who had been involved in the formal distribution of videos abandoned their work with the implementation of the ban. The same video clubs did not simply resume their operations as illegal institutions. Thus, in order to locate the newly underground infrastructure, we must trace the connections between the formal video industry and the underground infrastructure through equipment, protocols, and expertise.

During its four years of legal operation, the increasingly centralized video industry ensured that a range of home video technologies entered the country through both formal and informal means. These technologies included blank videocassettes, movies on video, video players, film-transferring equipment, and camcorders. The industry hadn't just provided the raw technological materials but also the professional and consumer literacy necessary to operate them. An entire professional union had spread technical information about videocassette technology to both technicians and the public. Meanwhile, easily accessible publications like *Cinema on Video* had provided detailed information about how to operate a range of home video equipment (figure 2.1). The growing rental industry had meant that video club employees and their customers were familiar with the technicalities of videocassettes and players. Because the ban came after the development of a video industry, technologies and knowledge of their protocols were already circulating widely in Iran.

راهنمای فیلم‌های مجاز ویدئو

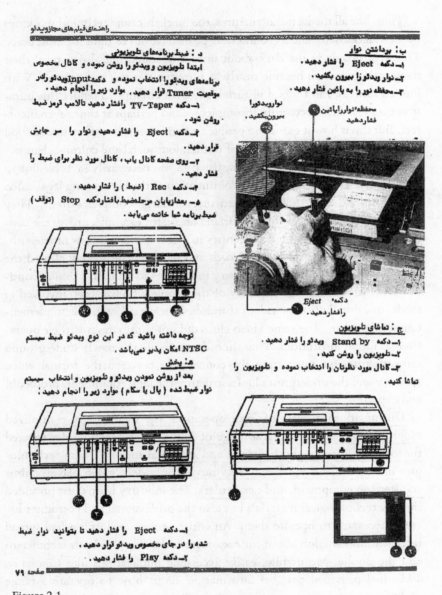

د : ضبط برنامه‌های تلویزیونی

ابتدا تلویزیون و ویدئو را روشن نموده و کانال مخصوص برنامه‌های ویدئو را انتخاب نموده و دکمه Input ویدئو را در موقعیت Tuner قرار دهید . موارد زیر را انجام دهید .

۱ـ دکمه TV-Taper را فشار دهید تا لامپ قرمز ضبط روشن شود .

۲ـ دکمه Eject را فشار دهید و نوار را سر جایش قرار دهید .

۳ـ روی صفحه کانال یاب ، کانال مورد نظر برای ضبط را فشار دهید .

۴ـ دکمه Rec (ضبط) را فشار دهید .

۵ـ بعد از پایان مرحله ضبط با فشار دکمه Stop (توقف) ضبط برنامه شما خاتمه می‌یابد .

ب : برداشتن نوار

۱ـ دکمه Eject را فشار دهید .

۲ـ نوار ویدئو را بیرون بکشید .

۳ـ محفظه نور را به پائین فشار دهید .

نوار ویدئو را بیرون بکشید

محفظه نوار را پائین فشار دهید

دکمه Eject را فشار دهید .

ج : تماشای تلویزیون

۱ـ دکمه Stand by ویدئو را فشار دهید .

۲ـ تلویزیون را روشن کنید .

۳ـ کانال مورد نظرتان را انتخاب نموده و تلویزیون را تماشا کنید .

توجه داشته باشید که در این نوع ویدئو ضبط سیستم NTSC امکان پذیر نمی‌باشد .

ه: پخش

بعد از روشن نمودن ویدئو و تلویزیون و انتخاب سیستم نوار ضبط شده (پال یا سکام) موارد زیر را انجام دهید :

۱ـ دکمه Eject را فشار دهید تا بتوانید نوار ضبط شده را در جای مخصوص ویدئو قرار دهید .

۲ـ دکمه Play را فشار دهید .

Figure 2.1
The legal video industry of the early 1980s supplied new users with knowledge about home video technology. Here the popular trade *Cinema on Video* details the anatomy of the video player.
Source: *Sinemā dar vide'o* 2 (October 1983): 79.

   Compared with the thriving rental industry before the ban, static spaces
dedicated to displaying, renting, and selling movies on video were rare during the ban period. They did, however, exist, even on a small scale—like the
small video club that opened this chapter. These physical stores catered to
cinephiles and other serious viewers. They featured higher-quality copies
of movies for rent and often a larger, more specialized art-house selection.
Private clubs had certain measures in place to protect them from the police
but also to ensure that customers returned the high-quality cassettes they
rented out. These measures included membership fees. As *Variety* reported
in 1984, membership at such a club ran upwards of $200 (approximately
$500 adjusted for inflation) per month in addition to rental fees.[10] These
stores also employed secret codes to stay hidden from the state's eyes. There
were, for example, two underground video clubs in central Tehran, both
of which were located at the end of narrow hallways and always appeared
closed. At one of the stores, potential customers would line up in the empty
hall. When it was their turn, they would knock on the door. The person
working at the club would answer and demand, "What do you want?"
Customers would reply, "What do you have?" The clerk would then say,
"I don't know. I have to go see." Then he would disappear and return with
two movies that the customer had to take. If customers did not follow this
script exactly, they would be turned away. Extreme examples such as this
show how the video store in Iran during the ban period was the exception
and not the rule. Because they were immobile, these spaces were vulnerable
to police raids and were, therefore, often short-lived.
   Following the ban there was increasing demand for the limited technology that existed in the country. Figures show, for example, that the cost
of a video player shot up from several hundred dollars before the ban to
the price of a new locally produced car after it was announced.[11] Addition-
ally, we know that at certain points the state banned the importation of
video equipment, which meant that the influx of video technology into
the country slowed. As a result, the informal video network that would
develop after the ban depended in large part on the same cassettes, players,
and devices that had sustained the legal video industry in the early 1980s.
The reuse of older equipment became a hallmark of this infrastructure.
We even saw it echoed in the story that began this chapter, as Simin spent
years searching for Amin's old videotapes among her video dealers' stock
of rental cassettes.

Despite these continuities, the underground video infrastructure wasn't the same system as formal video industry of the early 1980s—both in its procedures and in its reach. The underground market for videocassettes would ultimately far exceed the spread and financial value of the video industry that had mushroomed at the turn of the decade. Indeed, my corpus of oral history interviews reveals that most Iranians first encountered videocassette technology not through the legal video clubs of the early 1980s but rather through the underground system of video distribution that would emerge after 1983. This pattern of growth shows how an informal media system might exceed its formal antecedent. The wide reach of the underground infrastructure also enables us to reflect on how, despite limited equipment, the specific properties of video technology allowed for unprecedented growth, even under the extreme conditions of a ban.

The expansive underground video infrastructure in Iran was shaped by the material qualities of home video technology itself: plastic shells, magnetic ribbons, feeder wheels, knobs, and buttons. The rise of home video technology meant the materialization of motion pictures. No longer the ephemeral experience of the theater, movies suddenly became something that people could hold, touch, exchange, and even collect. As I will show in chapter 4, videocassettes transformed the home into a site of viewership. Although television had existed in Iran for decades, there was relatively little entertainment programming and it was mostly a news medium. The videocassette technology that became popular in the 1980s and 1990s turned the television set into an entertainment device. The fact that movies were now a material reality for the home drove much of the underground video distribution system.

Crucially, videocassettes were material objects that needed to be transported to the millions of individual users who demanded them. From trench coats and briefcases to wheelbarrows and laundry baskets, the modes of transportation that video dealers employed reveal a logic of invisibility that governed the Iranian video infrastructure in 1980s and 1990s. Along with basement workshops and back alleys, these modes of transportation were part of a larger system of concealment that marked how videocassettes lived and moved in the Islamic Republic. In some respects, this may not be surprising. As I've shown, scholars often describe the invisibility of infrastructure.[12] And yet infrastructures are not made invisible for the same reasons or to the same ends. Instead, the concealment of every infrastructure is socially, culturally, and sometimes ecologically conditioned—whether to prevent the public from advocating in policy decisions or to protect valuable equipment from the impact of natural disasters and shifting geopolitical alliances.[13] In Iran, the video infrastructure was necessarily invisible given its illegal status. Keeping it hidden left not only consumers in the dark but also the police. The viability of video distribution in the country depended on the invisibility of videocassettes.

Video dealers and their bodies were a crucial part of the underground infrastructure and operated at the threshold of visibility and invisibility. AbdouMaliq Simone has positioned "people as infrastructure." In his assessment, people can produce and distribute social services, especially in areas where public services have failed them.[14] This is a productive take on infrastructure, one that resonates with the underground video infrastructure in Iran. To extend this line, we might think about not only the processes that people forge but also how they use their bodies to implement them.[15] In Iran, the video dealer's body was a central mechanism in the complex machinery that transported videocassettes to consumers. With jail time and lashings as possible punishments for getting caught, it was the video dealer's body that was most vulnerable in the process of video distribution. The video infrastructure in Iran in the 1980s and 1990s did not just consist of plastic, metal, and cement pathways but also human bodies that negotiated and interacted with these elements.

For many people, the experience of movies on video became inseparable from the dealer's briefcase, which was full of cassettes. That small treasure trove was their access point to a whole world of sights and sounds. For the most part, everyday users of videocassettes were oblivious to the logistics of video production and transportation. The experience of movies on video began for the consumer with the arrival of a dealer and his collection of cassettes. At the same time, the video dealer as a figure was synonymous with the large briefcase he carried—almost always called a Samsonite in Persian regardless of its actual make. A video dealer could be old or young, tall or short. He could have long hair or short hair, have a beard or be clean-shaven. Only the briefcase he carried made him recognizable as a video dealer.

Over time the video dealer's trademark Samsonite made him vulnerable to getting caught. In her memoir *The Lonely War*, journalist Nazila Fathi recalls her family's video dealer in the mid-1980s. She writes, "Our former video-man Bijan . . . disappeared for several weeks. When he finally showed up, we found out he had been arrested. His boxy briefcase had given him away, since government forces on the hunt for video-men sometime stopped people with briefcases on the streets and searched their bags."[16] Such an account confirms just how expansive the underground video network really was. The video dealer with his Samsonite in tow was such a common sight on city streets that the police began targeting all men with a briefcase in hand. They hoped to find an illegal stock of videocassettes in transport, to discover an access point into the underground world of video.

The large briefcase may have been the standard way in which videocassettes travelled underground and between homes, but it was hardly the only means of transportation for movies on video. As the Samsonite became the dealer's trademark accessory and, therefore, his Achilles heel, video distributors had to resort to more creative options. During my interviews, I heard reports of laundry bundles, trash cans, and even wheelbarrows protecting stocks of videocassettes from view as they moved through public spaces. It was the trench coat, however, that emerged as the most practical replacement for the Samsonite.

As Fathi recalls in her memoir, when her family's video dealer couldn't reach them because of government roadblocks, he would send a colleague instead. This substitute, according to Fathi, "was a giant man who always wore a long trench coat. He never carried a briefcase, and he walked like a wrestler, his arms swinging back and forth inches away from his body. Once inside our home, he'd pull tapes from every pocket of his coat, nearly a dozen in all."[17] Of course, the trench coat wasn't just used to transport videocassettes; it was also used to conceal other forms of illegal media. Famously, Marjane Satrapi, in her graphic memoir *Persepolis*, remembers buying cassette tapes with illegal music from men who stood on street corners cloaked in trench coats.

As briefcases and trench coats kept videocassettes out of sight in Iran, they also connected media technologies to dealers' bodies. Pressed against their arms, chests, shoulders, and back in the folds of a coat, videocassettes were indistinguishable from video dealers and their bodies as they moved through public spaces. One former video dealer I spoke with remembered how sore his right arm and hand were after his first week delivering videocassettes. He was so nervous that first week that he had clenched the handle of his Samsonite with all his might. The extra strain left his right arm sore and exhausted. It was only with time, as he became more comfortable with the job, that he began to loosen his grip, playing with how tightly he needed to hold the briefcase to keep the cassettes in tow. At that point, he recalled, as his heavy briefcase dangled loosely from a few fingers, it felt like part of him. Statements like this affirm how a video dealer's body became entangled with the other material that constituted the underground video infrastructure.

Material forms of video technology—such as cassettes and players—constituted another crucial component of the underground infrastructure. The reusability of video technology was one of its main affordances and shaped what the underground video infrastructure would become. Cassettes could be watched, rewatched, taped over, and repurposed many times over. The way in which the video infrastructure in Iran depended on this feature was a result of the scarcity of video equipment in the country. While the demand for movies on video may have exceeded supply, it was only marginally. Thousands of video distributors ran successful small businesses devoted to supplying access to movies, while consumers were generally satisfied with the regularity of services and the diversity of options. This success was possible for two reasons: a rental market that capitalized on the reusability of analog video technology and a steady, if not limited, supply of both video equipment and content.

During the ban period, the underground video distribution network was primarily a rental market, which made the most financial sense. Just as in the United States, the exuberant cost of prerecorded content on video made owning cassettes too pricey for most people. Of course, in Iran the financial value of movies on video was not determined by media conglomerates like in the United States; rather, videocassettes were made valuable in Iran because they were scarce, illegal commodities. A rental system allowed underground video distributors to make the most of the limited video technology that existed in the country at the time, while also meeting the demands of a public hungry for movies on tape.

The rental market meant that customers only possessed a cassette for a limited period. They paid less and accessed a more diverse selection of movies, since they could rent far more movies than they could ever purchase. Video distributors benefited, too. They could continue making money off the same cassette with each rental. Its profitability was not limited to just one movie, either. When a movie had fallen out of favor on the market, the cassette could be taped over with new content. Although the quality of the tape decreased with each new movie, this way video distributors continued to make money off their valuable equipment, and consumers gained access to a diverse range of movies. Video rental fostered an invisible economy that could grow, even as the importation of video technology slowed—sometimes to a halt.

The logic of this rental market, especially its limited supply of cassettes and players, allows us to reevaluate the global history of home video technology. Most of the scholarship on video industries in the United States and Europe addresses the infamous format war between Betamax and Video Home System (VHS). In the 1970s, several different electronics companies raced to introduce analog video technology to the consumer market. While Sony's Betamax player would be the first video player to penetrate the market successfully in 1975, JVC quickly followed with its VHS format in 1976. The two systems were incompatible: a Betamax tape didn't work in a VHS player and vice-versa. At the same time, there wasn't room on the market for both formats, which provided essentially the same functionality. With higher picture and sound quality, Betamax was the superior technology. Nevertheless, by the mid-1980s Betamax's shares had dwindled considerably in the United States and Europe, and in 1988 Betamax only controlled 0.3 percent of the market.[18] The reasons for the VHS format's victory in the format war range from lower prices to longer cassette tapes to buy-in from other companies. The typical narrative is that within a decade Betamax's early success was eclipsed by VHS's dominance.

In Iran, however, the format war played out on an entirely different timeline. Betamax was the dominant format until the mid-1990s, until after the wholescale ban on video technology was lifted. Referred to as *navār-e kuchak* (little cassette), Betamax tapes sustained the rental market in Iran long after the victory of the VHS system in the American market. This made good sense. The supply of new cassettes and players into the country was irregular during the ban period. The video infrastructure was too precarious to introduce sweeping changes in the format. The entire system—including the material technologies and relevant expertise—had formed around Betamax cassettes. This pattern is consistent with what we know about the history of technology. As David Edgerton has shown, technologies persist around the world long after they have fallen out of favor in the United States and Europe.[19] In Iran, it wasn't until after the ban on video had been lifted that the VHS or *navār-e bozorg* (big cassette) would triumph over Betamax as the preferred format. After the ban, video technology could more easily enter the country and the Iranian market followed suit with global trends that favored VHS systems. The long life of Betamax in Iran provides an important corrective to work on video industries that emphasizes consumer behavior, legal discourse, and industrial strategies to understand the format wars. Social, cultural, and political forces—like those that constituted the video infrastructure in Iran—also shape a medium such as video.

As cassettes, both big and little, shuffled throughout the country, they fragmented movie culture. No longer limited to just a few film reels and theater screenings, the movies existed in millions of shapes and sizes. This fragmentation meant that movie culture could exist in multiple, and even contradictory, spaces at any given moment. We saw evidence of this in the story of Nahid and Behrooz in the previous chapter, as two government employees traded illicit movies in the same offices that were responsible for censoring media. The fragmentation that video infrastructure enabled also meant that *Terminator 2* might just as easily end up in the VCR of a secular middle-class family in northern Tehran as it would in Morteza Avini's private collection. This fragmented movie culture made individual, family, and community screenings all possible—sometimes in unexpected places. For example, in 1984 *Variety* reported on a showing of *Gandhi* (dir. Richard Attenborough, 1982) "on a medium-sized screen at $9 a head at one of the major Tehran hotels."[20] Ultimately, the decentralized nature of the video infrastructure allowed for increased access and new modes of engagement but also ensured its continued success. The same *Variety* article cautioned, "Vid clubs operate from offices only, and no street level homevid shops were visible."[21] The diffuse nature of the video infrastructure was necessary for its existence; it allowed it to remain underground.

The politics of invisibility that governed the video infrastructure could be found in other national contexts, as well. Across the Global South, analog technologies that allowed for easy reproduction fostered illegal and informal video markets. In Delhi, for example, the fragmentation of video distribution was materialized in the Parika Bazaar, a winding maze of tight, incense smoke-filled corridors, whose "notoriety came with the video boom."[22] Because it was a "free zone of piracy," it was often subjected to raids by anti-piracy enforcement agents.[23] During these raids, video marketeers took advantage of the bazaar's "labyrinthine structure" to evade, misdirect, and confuse detectives.[24] Just like in Iran, the video market in the Parika Bazaar of Delhi was only as valuable as it was fragmented. These informal video infrastructures were material and facilitated the movement of goods, but to survive they also had to disperse and disappear as quickly as they had appeared.

BLIPS, BLURS, AND INFRASTRUCTURAL PLEASURE

As a limited number of videocassettes circulated through the country, their technical features wore down and faltered. Every use taxed the videocassette's material properties. Transfers, pauses, and fast-forwards took their toll on the fragile magnetic tape that stored the magic of movies deep inside every videocassette. The cassettes that circulated through Iran far exceeded the intended number of views and copies. As a result, glitches, tracking, and white noise fossilized within the movies themselves. They did not interrupt movies so much as they became part of the movies. Over time, the ligatures caused by the video infrastructure became synonymous with it. Watching movies with distorted sights and sounds was simply part of the experience of accessing movies through the video infrastructure. As my interlocutors explained time and time again, they would watch a movie from start to finish even if it was hardly legible, even if the sound and image quality was so bad that they could barely follow what was happening.

Studying the circulation of videocassettes as a meaning-making process requires that we investigate how the material and technological forces that allow for the storage and transmission of information affect people's experiences, sensations, and interpretations. The underground video infrastructure in Iran produced an aesthetic of blips and blurs that embodied video's illegality and positioned viewers against the state. Although expansive, obscure, and diffuse, the infrastructure became visible in the glitches that marked the videos it transported. Constant use and excessive transference wore down the videocassette's delicate engineering. As a result, audiovisual distortions disrupted the movies that circulated underground. The aesthetic that the underground infrastructure produced embodied both the pleasures and the perils of video. On the one hand, the glitches marked a sense of escape and collectivity among viewers. On the other hand, these blips and blurs reminded people of the illegality of the video infrastructure and, thus, their participation in it was always already in opposition to the state.

Iranians did not necessarily know the specifics of how videocassettes circulated in the country, but they recognized that an underground infrastructure existed and left certain marks within the cassettes and on the movies they held. This was a sentiment that I heard often during my interviews. For example, a former video user named Javad told me, "We didn't mind the bad quality, and we certainly didn't blame the video dealer for it. After all, we knew where these videocassettes had come from, the journey they had taken, how they had arrived in our hands." When I pushed him further and asked where the videocassettes had come from, he was at a loss for words. He said, "Look, those cassettes were illegal. It was an underground market. We understood exactly that." The interruptions and glitches were par for the course when it came to accessing videocassettes in Iran, especially during the ban period.

For many Iranians, the material marks left by the underground infrastructure were also part of the pleasures of the video culture that flourished in Iran beginning in the 1980s. As families gathered to watch a new movie on video, sometimes figuring out what was happening on screen was like a game. People shouted and argued about the mise-en-scène and dialogue. As parents, cousins, and aunts and uncles fought over a word or some visual detail, the others in the room became the arbiters of peace. It encouraged a level of participation with the screen and with one another that was vastly different than watching a movie in a quiet, dark movie theater. Shirin, who was a teenager during the ban period, told me that sometimes a movie's sound and image quality was so bad that "all we could do was laugh." The emotions that were conditioned by a video's blips and blurs did not simply disappear. They trained an entire generation of viewers. Even today, when a streaming video buffers, out of the blue, Shirin remembers the feeling of squeezing between two of her uncles, the fourth person on a small three-person couch, as they all settled in to watch a new videocassette. For just a moment, before the video resumes, she feels happy and warm.

Shirin's memories point us to a history of what we might call *infrastructural pleasure* in Iran. The videocassette's degeneration produced a "bootleg aesthetic" of visual and sonic glitches.[25] These distortions were the material marks of the informal infrastructure that had produced them. They embodied not only the underground distribution network but also the pleasures inherent in accessing movies on video through that system. This infrastructural affect expands what we know about the glitches created when the videocassette's engineering is exceeded through informal practices like pirating. Video scholars have long been interested in the aesthetics of video degeneration and what they signal for viewers. For many of these scholars, the videocassette's technological excess intersects with the excess of intellectual copyright law. They have argued that in places like China, India, and the United States, the degradation of analog video—its glitches, blips, and blurs—announced a system of piracy that was contrary to the discourse on copyright. Intellectual property law represents cultural currency in an increasingly globalized world; to subvert, overlook, or otherwise ignore copyright law in piracy is to use the technologies of globalization to reject or even critique the neoliberal policies that undergird it.[26] Thus, the reasoning goes that videocassette's audiovisual distortions reified an informal media economy and its sometimes contentious relationship with intellectual property law.

While copyright law organizes most of the scholarship on informal media practices, the underground network in Iran offers an opportunity to think about how video distortions also embodied local political negotiations and not just a global discourse like intellectual property. After all, copyright was not a serious public concern in Iran until the early 2000s, intersecting with the rise of digital rather than analog technologies.[27] Because Iran has never been a signatory to international copyright agreements, video users, especially those of the 1980s and 1990s, had a different relationship with the illegality of their cassettes. In Iran, it had less to do with the interests of big media conglomerates and more with the local policies dictating media consumption. In other words, what made videocassettes illegal—and therefore pleasurable—was the law of the land. Ultimately, the distortions that marked the videocassettes of the underground network demand a framework for informal media practices that moves beyond copyright, intellectual property, and globalization.

The infrastructural pleasure produced by the underground network's video degeneration registered a particular political subjectivity among viewers. That pleasure both departs from the existing scholarship's focus on copyright and affirms theories about the structures of feeling that media access can engender. As Larkin argues, the distorted sounds and images that mark informally distributed videocassettes evoke an infrastructural imaginary for many people. Within this imaginary, people not only assess their own modes of media distribution and access but also imagine what infrastructure should or could be. Writing of Nigeria, he suggests that video distortions "evoke fantasy of other countries where deficiencies in infrastructure are believed not to exist."[28] This is a useful starting point for thinking through the glitches that indexed the underground network's infrastructural pleasure in Iran. In Larkin's scheme, the conditions of urban Nigeria presuppose that infrastructures are inherently good or bad—successes or failures within a global system of infrastructures—and that people can distinguish between them. Video glitches may have been technological failures, but for Iranians in the 1980s and 1990s, they did not embody the failure of the underground network but rather its successes. The underground network was always a success because it created access to moving images where access wouldn't have otherwise existed.

For this reason, it is generative to think about analog video distortions in terms of an "aesthetic of access."[29] As Hilderbrand suggests, such a "bootleg aesthetic . . . exhibits the audience's engagement in a clandestine love affair—watching, sharing, and copying the illicit text."[30] He argues that the videocassette's audiovisual warps incited and excited bootleg culture in the United States, as the pleasures of home video technology interlaced with the subversion of copyright law. Although copyright law was not central to the experience of the underground network, there are certain parallels we can draw to Iran in the 1980s and 1990s, where the bootleg aesthetic was also the materialization of an underground "love affair" with movies. Hilderbrand writes, "Materially, the fall-out of the image and soundtrack mark each successive copy as an illicit object, a forbidden pleasure that has been watched and shared and loved to exhaustion."[31] The videotape in Iran similarly captured a relationship of pleasure and illicitness through its audiovisual distortions.

Yet one must resist the urge to fetishize this pleasure. As Laikwan Pang has cautioned, scholars should be careful not to assess this bootleg aesthetic through a rubric of liberal, leftist politics. She writes of China's ascent into a global, neoliberal economy in the 2000s, when intellectual property law suddenly mattered there in unprecedented ways. She argues that while bootlegging and its subsequent aesthetic appear to subvert regimes of intellectual property, they are also almost always driven by a desire for the commodity itself.[32] Thus, bootlegging in China is perhaps less radical than it may appear, since it is driven more by the logic of consumerism than by resistance to copyright law. We might expand this assessment to understand the circulation of videocassettes in Iran in the 1980s and 1990s, as well. Iranian video users prioritized accessing movies and didn't necessarily prioritize breaking the law. Nevertheless, the glitches that marked these movies directed viewers to the very mechanisms that had provided their access in the first place. Those mechanisms were undeniably illicit and, thus, the pleasure of the movies became subsumed in the illicitness of their accessibility. Comparisons like these between Iran, China, and the United States direct us to a more general theory of video infrastructure, one that exceeds the particularities of a single national context.

As one former user, Behnam, told me, "We rented videocassettes because we wanted to watch movies. The fact that it annoyed the government was just the icing on the cake [*nur-e 'alā nur*]." "In fact," he continued, "sometimes the movies were so messed up, all we could see was *light*." He laughed. His joke played with the turn of phrase *nur-e 'alā nur*, which literally means "light upon light." In his joke, Behnam draws on the idea of light pouring from light to signal the washed-out images that were so familiar to video consumers at the time. He explained, "Black-and-white movies like *Qeysar* [1969] or *Casablanca* [1942] were more white than black. As a kid, I remember sitting in a dark room that was glowing with the movie's white hues. With so much light, we could barely discern the shapes of the characters. We loved it anyway and walked like Qeysar for weeks afterwards." In Behnam's pun and his further explanation, we can see the triangulation of pleasure, illicitness, and the aesthetics of access as they became entangled, embedded in, and circulated through videocassette technology.

The aesthetic of blips and blurs gave form—however warped and unstable—to the infrastructure that had supplied its users with access to movies on video. Scholars such as Star have argued that infrastructures are "by definition invisible."[33] The videocassette infrastructure in Iran, however, asks that we soften that definition. As we saw in the previous section, the underground infrastructure was dispersed, and its mechanics were often concealed for users. Yet that invisibility was never absolute. The glitches that blotted many underground videos were visible marks that allowed viewers to imagine an underground system of distribution. The video infrastructure thus operated at the nexus of visibility and invisibility.

The kind of infrastructural intelligibility afforded by the glitches of underground video distribution fostered a sense of community and collectivity among its users. The blips and blurs embodied video's illegality as much as the pleasure of participating in a public secret. Behind closed doors and within private spaces, users felt connected to one another, even strangers, through their awareness that many other people were engaged in the illicit rental of videocassettes. This collective identity was made possible through the material endpoints of the video infrastructure. A video dealer's body, a videocassette's shell and ribbon, and the occasional brick-and-mortar store materialized the video infrastructure for users, if only temporarily. These moments of materialization—when users became aware not only of the infrastructure itself but also of the other people who engaged with it—allowed them to conceive of a different collective cultural identity than the one put forth by the state. This was an underground network, in which the materiality of videocassettes, as much as any content that appeared on them, created the sense of shared culture and identity.

The myth of the mother cassette was one embodiment of this sense of connectivity. While most of the people I interviewed, including dealers, confessed that they had never seen a movie on video in its original packaging, they nevertheless waxed poetically about the mother cassette (navār-e mādar) that must have spawned all subsequent versions. In this understanding, the mother cassette was the studio-manufactured cassette that entered the country illegally, perhaps in the suitcase of some traveler abroad. The mother cassette was the start of it all. From there, dealers and other laborers reproduced it en masse and distributed it around the country. While hundreds or thousands of copies of the same movie might be on different cassettes, they all had one point of origin. As Hilderbrand has noted of bootleg culture in the United States, "the proliferation of degenerate copies . . . contributes to the allure of an idealized filmic original."[34] Similarly, the mother cassette in Iran was a device that allowed people to imagine a network of videocassettes that was too diffuse and too obscured to be seen or fully understood. It deployed the image of family—of mothers—to convey the sense of connectivity that was essential to the underground video network.

In addition to the videocassettes themselves, dealers were the other main point through which urban Iranians encountered media infrastructure in the 1980s and 1990s. As a result, through the video dealer, people came to understand the expansiveness of the underground network. These men and (at least in one instance) women embodied a complex system for the circulation of videocassettes while also representing a community of people behind it. In the most basic sense, video dealers were other people who had other customers. Even without meeting these other customers, users were aware of their existence. While the official discourse on video dealers positioned them as hardened criminals, in practice most people recalled their video dealers with "fondness."[35] Dealers were the main point through which urban Iranians encountered media infrastructure in the 1980s and 1990s. During my interview with Arezoo, who was a child in the 1980s, I asked if she remembered the name of her video dealer. "We called him 'Amu," she recalled, and chuckled. "'Amu" is the Persian word for paternal uncle, but it is also a way that children often address men who are close to the family but not necessarily related. Arezoo reflected further, "You know, he was like an uncle. He connected us to an entire secret family of other video watchers." Her assessment shows how video dealers reminded people that they were part of a larger illicit community—perhaps even a family—of videocassette consumers.

Throughout the 1980s and 1990s, video rental was a secret that everyone knew but no one talked about. During this time, the state fostered a culture of surveillance, pitting neighbors against one another and encouraging them to report illegal activities, including watching movies on video. As Nahid, a former employee of Islamic Republic of Iran Broadcasting (IRIB), explained:

> The people who would bring the films . . . had to be cautious. They had to be careful not to be noticed, and sometimes they would come late. We would be home, waiting for them to come. I always thought that they had certainly gotten into trouble, because they were late, or they wouldn't come for two or three days, and you wouldn't know where they were. We lived in an apartment building, and . . . the people who distributed films would come to people's houses. For me, it was always interesting to imagine what the neighbors thought, when once a week, in the morning or afternoon, someone would come to our house and after five minutes, exit and leave. This has always been one of those questions that I still don't have answer for. [Laughs]

In this testimony, concepts like trouble, getting caught, and arriving late characterize Nahid's encounter with the video dealer. Also implicit in this statement is the tension between the visibility and invisibility of infrastructure. Her concerns at the end of this statement attest to the realities of urban life. In densely populated cities, people are acutely aware of their neighbors. This was especially the case in the 1980s as the government called on citizens to report their neighbors' criminal activities.[36] Nahid's laughter as she imagines what her neighbors must have thought about the quick arrival and departure of her video dealer captures her ambivalence. Video dealers were supposed to remain invisible, so their visibility in the apartment building heightened Nahid's awareness of the process of video distribution, the illegality of the dealers' work, the risks they took, and the precariousness of her own position as a consumer of these videocassettes.

The video infrastructure was public and private in the same moment. It ushered cassettes through public spaces, where they fostered a sense of togetherness. Videocassettes usually ended up in domestic spaces, like Nahid's apartment. And yet those spaces were not entirely private, as nosey neighbors kept tabs on one another. The video infrastructure, through its procedures, routes, and materials, helped produce and even institutionalize the underground. Iranians of the 1980s and 1990s increasingly occupied that space, where the boundaries between public and private were not entirely clear and yet mattered a great deal. In its first two decades, the Islamic Republic encroached on private spaces, insisting that they be regulatable like public spaces. The state sought to control people's private lives, including the media that they consumed. Nevertheless, people led private lives that they kept sheltered from the state's watchful eye, and these private lives included movies on video. Through their engagement with the video infrastructure, Iranians negotiated the bounds of regulation, law, and enforcement under the young Islamic Republic.

These negotiations were active processes. We can see evidence of that in interactions between the video dealer and his customers. As I argued in the previous section, the dealer's body was a constitutive part of the underground video infrastructure in Iran. That body was necessarily gendered. Overwhelmingly, video dealers were men. As a gendered body, it challenged certain protocols. In Nahid's earlier statement, the punchline revolves around the fact that a strange man is entering a woman's home. Her laughter punctuates the sexual innuendo, as does her question about what the neighbors would think. Here it isn't just about what is legal or illegal in the Islamic Republic but also about what is proper for a woman in Iranian culture.

Elsewhere, too, I encountered evidence of how these gendered encounters between dealer and customer reflected the process of negotiating public and private, legal and illegal. As one interviewee would tell me, his mother would cover her hair every time their video dealer stopped by with a fresh collection of cassettes. This wouldn't necessarily be strange in Iran, except that his mother was not religious and, in fact, adamantly opposed the mandatory veil. She normally would not wear a headscarf when a male visitor came over. Such an anecdote suggests a mixing of codes. On the one hand, his mother would access the illegal world of videocassettes. On the other hand, she veiled herself according to the law of the land. This paradox raises questions. Did the video dealer's body, as a part of the circulation of videocassettes, turn the home into a public space that demanded veiling? Or was the mother trying to take the edge off the illegality of video by otherwise performing the law? What is clear in an interaction like this is that by accessing videocassettes and tapping into the video infrastructure, individuals operated on the periphery of what was legal, and they began to develop the strategies, protocols, and ethics necessary to occupy such an underground space.

To view the video infrastructure simply in terms of its endpoint—the moment when the videocassette was slipped into the VCR—is to overlook the complex technical and cultural procedures that institutionalized movie culture in Iran, however informally. Videocassettes did not just exist in private homes. Whether they originated in the video dealer's workshop or in one of the rare video clubs, they also had public lives. The very point of the video infrastructure was the safe delivery of videocassettes through public spaces and to people's homes. As Larkin has argued, infrastructures, by facilitating the movement of material goods, gather people into "collectives."[37] In Iran, that sense of connectivity pervaded the video infrastructure. People's engagement with and understanding of the video infrastructure took them far outside of the safety of their own homes.

Today's networked society is often posited as a contemporary phenomenon, the result of recent digital technologies. Yet historians have shown that systems of media distribution have long fostered a feeling of togetherness—from libraries and reading publics in the eighteenth century to telegraphy as the "Victorian internet" in the nineteenth century.[38] The sense of togetherness nurtured by media, though, is not an inevitability. Rather, it is mediated by the social and cultural conditions of a particular time and place. In Iran in the 1980s and 1990s, the underground video infrastructure fostered a collective identity because the people who participated in it were aware of its illegality. This knowledge encouraged an illicit community of users, who felt intimacy with one another through their pleasurable engagement with a banned medium and the infrastructure that ensured its safe delivery.

ACROSS BORDERS AND OVER SEAS

While the particularities of Iran's social, cultural, and political context may have conditioned the meaning of video for its many users, the movement of videocassettes within the country also depended on vast transnational networks. Given the ban, these informal, transnational channels were important to the spread of video in Iran. The video infrastructure relied on some equipment from before the ban, but new cassettes and, importantly, new content were also necessary to support its growth. People recognized videos as global commodities. They imagined the faraway places and complicated transnational routes that had allowed them access to movies on video. These imaginaries were important to how people positioned themselves in relation to the state and the world. The underground distribution of videocassettes—which people imagined as embedded in the global circulation of goods—offered a cosmopolitan subjectivity that contrasted with the isolationist national identity that the government sought to impose.[39]

Although all video technology was officially banned, it nevertheless made its way into the country through a multitude of formal and informal channels. In line with the law of invisibility that determined the underground video infrastructure, none of the video dealers with whom I spoke knew exactly where their supply of movies on video had originated. This shows how the infrastructure was so obscured that even the laborers embedded within it did not fully understand how it functioned. Nevertheless, there is some evidence that guides us toward the supply networks that fueled video distribution. With no videocassette factories in the country, the underground market depended on lax airport customs procedures, porous borders, and the reintroduction of confiscated material into the black market.

Airports were important points of entry for new videocassettes and content. In 1986 the trade publication *Film* published a brief report about a new ordinance that the Ministry of Islamic Culture and Guidance (MCIG) had released, instructing airport customs officials to confiscate all videocassettes and send them to the Ministry for inspection. According to the ordinance, the reason for this policy was that cassettes with prerecorded content were being "masterfully" repackaged to look like unopened (ākband) blank videocassettes. The ordinance concludes with a reminder that "given the prohibition of all video-related activities, the entry of blank cassettes on one's person is illegal." The title of the report in *Film* emphasizes this final point: "The Importation of Blank Tapes Is Now Banned" ("*Vorud-e navār-e khām mamnu' shod*").[40] While it would be easy to read this document as an affirmation of the state's definitive policy on video, a closer look reveals a much more complicated set of circumstances. Published three years after the ban on video technology went into effect, the report does not simply restate the illegality of video-related activities; it clarifies what exactly it meant. In other words, the initial ban on videocassettes hadn't necessarily precluded the importation of blank cassettes through airports—or at least not with enforcement. On a practical level, this meant that for three years, until the MCIG issued this ordinance, people were bringing blank cassettes into the country in their suitcases and evidently smuggling prerecorded content, as well.

Despite rigid instructions to confiscate illegal items, uneven implementation meant that travelers from abroad could make it through airport customs with their stock of contraband items intact, carefully hidden between the folds of their clothes.[41] After the Iran-Iraq War, members of the diaspora traveled back to Iran to visit family with more frequency. Despite the ban, they often tucked away in their suitcases a stash of Disney movies, which were gifts for relatives and also entertainment for their own children traveling with them. I spoke with some of the children of people who smuggled cassettes; they remembered the horror of watching their beloved copies of *Fantasia* (1940) and *Sleeping Beauty* (1959) being taken away. Now, as adults, they are convinced that the customs agents confiscated the Disney movies with the intention of bringing them home to their own children.

Despite these incidents, the very fact that travelers continued to bring cassettes with them from abroad shows how the airport remained an important space for the arrival of new video equipment. For example, I spoke to Shahryar, who was a child in Iran in the early 1990s. He recalled his father returning from the United States in 1993. He had promised to bring Shahryar the newest Disney release, *Aladdin* (1992), on video. Instead he arrived empty handed—much to Shahryar's disappointment. As his father explained, airport officials had searched his bags and confiscated the cassette. They claimed that Jasmine's outfit on the cover was inappropriate. This was just a pretense, however. Shahryar's father returned to the airport several times to plead with customs agents. After he paid off the right people, they released the video, even though the ban was still technically in effect. *Aladdin* would become one of Shahryar's favorite movies and the videocassette one of his most prized possessions. He remembered carefully slipping it under his shirt and taking it to his cousins' house down the street.

Airports were not the only port of entry into the country for videocassettes. They also entered the country across land and sea borders. For more than a millennium, Persia's many borders have connected it to Anatolia, the Caucasus, the Indian Subcontinent, and the Arab world. With historical precedents like the Silk Road, many of these connections were forged through trade and the movement of material goods. It is not surprising then that videocassettes carried on this tradition in the 1980s and 1990s, becoming part of a larger tradition of smuggling. I spoke with former video users in Baluchistan, close to the border with Pakistan, and Bushehr, a city on the shores of the Persian Gulf. Unlike their compatriots in the capital city, they remembered a steady influx of new videocassettes and new movies on video. They even owned a few cassettes of their own without having to rely on a rental system. The selection of movies available to them, which included large numbers of Hindi films, helps trace the movement of video technology through southwestern borders of Iran. Being on the periphery of the country allowed freer access to videocassettes, both because of the proximity to smuggling routes but also because of the distance from the political and cultural capital, where enforcement of laws tended to be more concentrated.

These accounts show how the underground infrastructure was embedded in a global circuit of goods, which neither started nor ended in Iran. As *Variety* reported in the 1980s, a steady supply of movies on video entered Iran from the United Kingdom. According to one article, "Most of the tapes, coming from England, are pirated and moved into Iran by an Indian ring."[42] This report reveals the global reach of video at the time, while also demonstrating how, perhaps unexpectedly, colonial networks between the UK and India helped determine the circulation of videocassettes in Iran. Ultimately, the underground infrastructure in Iran was one nodal point in a vast global network of video distribution. In the early 1980s, *Variety* reported that video equipment traveled through Iran en route to the United Arab Emirates. According to the report, the Iranian market absorbed 60 percent of the shipments.[43] Similarly, *Variety* claims that the underground distribution system in Iran supplied movies on video to Pakistan, where videocassettes and players were also illegal at the time.[44] Of course, these reports have an axe to grind. The trade publication represents the interests of Hollywood studios and, therefore, vilified informal systems of media distribution. Nevertheless, these accounts gesture toward the global expanse of video at the time and the ways in which the underground video infrastructure in Iran was part of a greater transnational circuit.

The global circulation of videocassettes affected how people imagined and understood the underground video infrastructure. As I showed in the previous section, videocassettes connected users to an illicit community within Iran. But the way in which they imagined the transnational reach of video also connected them to the world. In my interviews, I asked my interlocutors where they thought new movies came from during the videocassette period. All of them pointed me to various ports and borders. One person speculated that most new material must come through Turkey because the western borders were tied up with drug smuggling. Another person thought that videocassettes must enter the country along with drugs from Afghanistan. Others conjured that they must come through ports on the Persian Gulf or the Caspian Sea because they were already set up to receive shipments and contraband items and could more easily be slipped in with other arrivals.

Whether by land, air, or sea, the fact that people imagined videocassettes traversing Iran's many ports and borders shows how they envisioned them as a small part of a much larger world. In tandem with my interlocutors' speculations, in her best-selling memoir *Reading Lolita in Tehran: A Memoir in Books*, Azar Nafisi claims that, according to popular legend, in the 1980s sailors on the Gulf had permission to watch forbidden videocassettes, and they were responsible for smuggling them into the country where they circulated uncontrollably.[45] Where the cassettes actually came from is less important than the fact that people saw them as originating in borders and ports. This shows how they understood them to be global commodities. The way in which everyday people encouraged and supported the circulation of videocassettes showed their desire to be part of this global culture.

These were small transgressions that resisted the here and now and the call to nationalism that sang across other media. At a time when a new government and a war insisted that people consider what it meant to be good Iranians, the millions of cassettes that swirled through Iran in the 1980s and 1990s supplied tiny pieces of somewhere else.

Movies on video offered a view to the world that was otherwise shut off for many Iranians. It wasn't just that videocassettes offered an escape from the official state narrative; they also allowed people to experience other places and sometimes other worlds entirely. *Star Wars: Episode IV—A New Hope* (dir. George Lucas, 1977) was one of the movies that almost everyone I spoke to remembered watching on videocassette. Omid, whose family had a strict rule that they did not rent war movies, always made an exception for *Star Wars*. It was one of the few movies they rented multiple times, until it fell out of circulation with their dealer in the late 1980s. Omid remembered that the version of *Star Wars* he grew up with was impressively dubbed. But for some reason the film's first and most famous intertitle—"A long time ago in a galaxy far, far away"—had been left untranslated. As the best student (*shāgard-e avval*) in his English class, it fell on young Omid to translate the movie's first line the first time the family watched it together (after rewinding and pausing, of course). His pocket English-Persian dictionary, which didn't have the word galaxy, failed him, so he turned to his family's large reference dictionary. Once he had finally nailed down a suitable translation and read it aloud, he remembers his parents sighing. Their sighs echoed Omid's own feelings of relief that for the next two hours, they would be transported somewhere far, far away.

The oft-quoted first line of *Star Wars* represented more than the start of a family favorite; it was emblematic of a larger desire to escape from the here and now. This made sense given the violent war that had taken over the country. Sociologist Abbas Kazemi, in his own assessment of the videocassette period, explains that during the 1980s Iranians operated between two dominant modes: "living in hardship" and "living in distress."[46] While hardship represented the material struggles of a wartime economy, including the lack of food, distress was the psychological response to the hardships of war. Caught between these two pulls, people embraced videocassettes, in Kazemi's estimation, because "video was a medium of leisure in a society without leisure."[47] As Fathi describes in her memoir, the early 1980s witnessed a flood of Indian movies on the underground market. She writes that the colorful, musical scenes that populated Indian movies were a "welcome distraction from the bloody images of war on state television."[48]

The selection of movies available through the underground reinforced the cosmopolitan identity that the video infrastructure nurtured. When I spoke to video dealers about their stock of movies, they all insisted that they rented primarily Hollywood films, alongside a spattering of old Iranian films, occasionally Indian movies, and more rarely selections from European arthouse cinema. Their customers, however, recalled a very different selection of movies. As one essayist reported, "the most important movies that we watched on video, other than popular Iranian movies from before the revolution [*filmfārsi-hā*], were *Rocky, Sholay*, Bruce Lee's *Enter the Dragon* and *Fist of Fury*, and things like that."[49] Such a global list, which casually mentions popular titles from the United States, India, and Hong Kong, corresponds to the film selections that I heard from the people I interviewed. With just a few titles, they managed to cover some of the biggest film industries in the world, crisscrossing North America, Europe, and East and South Asia. With these samples, it wasn't so much that my interlocutors tried to deny the dominance of Hollywood as much as they viewed videocassettes as representing a broad swath of world cinema, from Italy to Armenia to Serbia and Korea.

With a steady supply of movies from all over the world, the generation of Iranians who grew up on video delighted in the global film pedagogy that the underground infrastructure provided them. During my interviews, I spoke with several former videocassette users who studied cinema in Europe or the United States. They all agreed that compared to their European and American classmates they had seen more movies. But it wasn't just a matter of quantity. It also had to do with geographic cover. They could just as easily discuss *Gone with the Wind* (dir. Victor Flemming, 1939) with their American classmates as *Sangam* (dir. Raj Kapoor, 1964) with their Indian peers. According to these interlocutors, their film education on video had been a global curriculum that didn't necessarily distinguish between the hierarchies that normally dictate the global marketplace. People watched what was available, which included examples of world cinema in addition to the staples of Hollywood and Iranian movies. Because what was available for rent on video was almost always separate from the Islamic Republic's state-controlled media, it always represented some other time and some other place entirely.

The transnational quality of the underground video infrastructure mattered because throughout most of the 1980s and 1990s, Iranians experienced a growing sense of isolation. The revolution and the establishment of the Islamic Republic restructured Iran's foreign interests and diplomatic ties. The international community's condemnation of the semi-sanctioned attack on the American Embassy left Iran with few allies in the early 1980s. The country became further isolated when Iraq attacked its southwest border in the fall of 1980. The war created a rhetoric in the country that pitted Iran against the rest of the world, especially since most global powers supported Iraq in the conflict. At various points throughout the war, the government closed the borders entirely, so that travel in and out of the country became temporarily impossible. The revolution and the war also meant that many urban Iranians were disconnected from friends and family living abroad. Even after the war, Akbar Hashemi Rafsanjani's privatization policies focused on internally repairing Iran rather than on engaging the global economy. Not until Mohammad Khatami was elected president in 1997 did an Iranian leader actively seek to break Iran's isolationist policies and engage the international community.

As Iran became more isolated from the international community, many Iranians also felt disconnected from the national-religious identity that the new government had determined for them. This sanctioned subjectivity was inescapable, since it was at the core of most legally circulating media content at the time. Television and radio played an important role in transmitting ideas about the war to everyday people. This was accomplished through the rhetoric of the state, which emphasized the war either as the Sacred Defense (*defā'-e moqaddas*) or the Imposed War (*jang-e tahmili*). The state's religious rhetoric did not resonate with most of the middle- and upper-class urban Iranians I spoke with. These designations cast the war as defensive and religious and, thus, heightened Shi'a nationalist sentiment, which left many people feeling unaccounted for within the nation. In the face of these forces of isolation, the underground video infrastructure offered something else: it provided a global supply of movies that allowed viewers to see and experience other places. This cosmopolitan identity did not just depend on what people watched but also how they accessed it. Videocassettes were material objects that needed to be transported through transnational networks to arrive in Iran. Users imagined these networks, their routes, and ports of entry into the country. In doing so, they viewed themselves as global citizens, as part of a worldwide community of video consumers.

LAYERS, MATERIAL, INFRASTRUCTURE

Even as the analog technologies of videocassettes have given way to digital media, the underground infrastructure in Iran continues to operate as a collective of material, bodies, and protocols. These days media distribution seems to exist in the clouds. Over the last decade, the advent of digital technologies has meant that viewers around the world can stream movies directly into their homes without having to shuffle around cassettes or discs. No longer bound to the theater walls, movie culture has become immaterial—or at least that's what popular discourse would have us believe. Critical media infrastructure scholars have disrupted this narrative by showing that even today's digital media are grounded in a material reality through cables, data centers, and antennas. As is often the case with infrastructure, the materials of media distribution overlay one another so that fiber optic cables, for example, now exist in the same channels that once supported transatlantic telegraphy cables.[50] In Iran, we see a similar process of infrastructural layering. The underground infrastructure forged through plastic videocassettes still functions as an important system for media distribution. It is still material and offers a collective identity that exceeds both state control and the country's borders.

We can see parts of the modern-day underground infrastructure in director Jafar Panahi's film *Taxi* (2015). The film was hailed by international critics as a "subversive piece of underground filmmaking."[51] It excavates the infrastructural layering that I have described while also demonstrating how the underground continues to operate as a space where alternative national subjectivities are possible. Panahi's successful run of feature films was interrupted in 2010, when the Revolutionary Court banned him from filmmaking because of his support of the protests after the controversial reelection of Mahmoud Ahmadinejad in 2009. Since then, he has actively defied that ban and continues to make movies that find success internationally. In *Taxi* the director drives around Tehran in a bright yellow cab, picking up passengers, listening to their stories, and secretly documenting his encounters. As is typical of Panahi's filmmaking, *Taxi* plays with the boundaries between fact and fiction. In some moments, the director's conversations with his passengers appear candid and unscripted; in others he reveals the artifice of his filmmaking through impossible coincidences and bizarre role reversals. Even his passengers sometimes seem uncertain about whether or not their encounters have been staged.

One of the passengers early in the film is a man named Omid, whom Panahi does not immediately recognize despite his insistence that they know each other. He eventually announces himself as Omid the Filmi. As it turns out, Omid used to bring movies to Panahi's house. Although Omid supplied movies to Panahi's son, Panahi was there one time and requested *Once Upon a Time in Anatolia* (dir. Nur Blige Ceylan, 2011) and *Midnight in Paris* (dir. Woody Allen, 2011). We learn that Omid specializes in the rare and the obscure, although he also carries stocks of *Big Bang Theory* and zombie movies to appease his customers. In the taxi, he offers Panahi a DVD, which Panahi notes hasn't been released yet. This is nothing, Omid claims. He can get dailies of films that are still in production. Omid's knowledge of cinema is as vast as his enthusiasm. He even calls Panahi out on camera, declaring "You're filming, aren't you?" He continues, "Want to know how I figured it out? The man's last line was very similar to the coffee scene in *Crimson Gold*." Here he is referencing Panahi's 2003 film *Talā-ye sorkh* (Crimson gold). *Crimson Gold* was hugely successful internationally, and even won the Jury Award at the Cannes Film Festival, but it was never released in Iranian theaters. The film's Iranian viewership, therefore, depended on its illegal circulation on videocassette and DVD. This moment captures both Omid's impressive knowledge of cinema—especially his ability to pick up on the intertextual clues that Panahi loves—and the centrality of the video dealer as someone who circulates movies so that they can find a life among Iranian viewers.

Once it has been established that both men are in the movie business, Panahi drives Omid the Filmi to one of his deals. A filmmaking student comes to the car and begins looking through Omid's collection, tucked into a large canvas bag. When someone calls Panahi's phone to talk to Omid, he is pulled out of the frame, and Panahi begins advising the student on what movies to buy. In a gesture that has become typical of Panahi's recent work, the roles are thereby reversed. Panahi becomes the video dealer rather than the filmmaker. The student confesses that Omid has told him that he and Panahi are colleagues (*hamkār*), and in that moment it makes sense, as Panahi performs the work of the video dealer by offering his suggestions and handing him a stack of DVDs in plastic sleeves.

In Panahi's films, these attempts to blur social boundaries are never meaningless. As Omid has an important business conversation on Panahi's phone and Panahi takes over the work of distributing DVDs, the hierarchy between production and distribution ceases to exist, and so too does the distinction between formal workers in the film industry and the informal laborers who do important work circulating movies. These hierarchies and distinctions matter to Panahi. After all, he has been banned from officially making movies. He continues to direct informally in Iran, although with the support of international distributers. Both he and Omid have an important role to play in terms of fostering a movie culture in Iran.

Taxi affords what Parks calls "infrastructural legibility"[52] by allowing us to imagine the underground infrastructure as it operates in contemporary Iran. Notably, the film directs our attention to the materiality of media distribution. With a taxicab as its framing device, mobility is a central motif in *Taxi*. People move throughout the city but so do media. Movies are material goods that need to be shuffled, dealt, and transported. In Iran, censorship, slow internet speeds, and limited bandwidth make streaming as difficult as it is expensive. DVDs, therefore, continue to play an important role in how people access cinema. Video dealers now work with DVDs or with flash drives so that people do not have to use their own internet account to download and stream movies. The underground infrastructure also continues to blur distinctions between formal and informal media practices. Just as the underground video distribution system preserved movie culture in the 1980s, the contemporary video distribution system equips users with access to Panahi's banned films, while also keeping film students stocked with a wide range of world cinema—from Turkish auteurs like Nur Bilge Ceylan to notable American directors like Woody Allen. As movies and series from around the world circulate on DVD and flash drives, they bundle people together, often in unexpected configurations, as Omid the Filmi ends up in Panahi's taxicab film studio. And yet their pairing makes sense, in that both operate against state attempts to control media.

The critical reception of *Taxi* overlooks the generative possibilities of the underground video infrastructure by simplifying it to a pirating network. This is especially apparent in its dismissal of Omid. Reviews of the film in the United States and Europe were overwhelmingly positive, calling the film "remarkable" and Panahi "admirable."[53] At the same time, Omid has been reduced to labels like "pirate" and "bootlegger" in these accounts.

In his review of the film in the *Guardian*, Jonathan Romney does give meaning to Omid's work as the "hero of samizdat distribution," but in the same sentence calls him a "bumptious clown."[54] In these kinds of accounts, the video dealer may have a role to play, especially within the context of the Islamic Republic's exclusionary policies, but they embody a system that shouldn't be taken too seriously. By calling Omid a "pirate," these reviews connect him to global debates about copyright that are taking place right now and are controlled by large media conglomerates.[55]

What this critical reception of *Taxi* refuses to acknowledge is the underground infrastructure as a site that brings people together and allows them to enact subjectivities other than what the state has determined for them. Such shortsighted reviews overlook the details of the film itself. As Omid exits Panahi's taxi and tries to pay, the famous director tells him there is no fare and to save his money for his "cultural work" (*kār-hā-ye farhangi*). This is precisely the point. Panahi, who also operates on the outskirts of what is legal in Iran, understands what it means to work informally. He is also aware of the history of the video dealer. Omid as a modern-day video dealer isn't just part of a recent global explosion of piracy; he's the inheritor of a long tradition of informal laborers providing Iranians with access to movies on video. Far from being bumptious clowns, the video dealers who began working in the 1980s were a central part of a vast underground infrastructure who helped to shape entire generations of media consumers.

VIDEO DEALERS: THE WORK OF INFORMAL
MEDIA DISTRIBUTION

Hamid waited patiently while the agents shuffled through his papers. Jeans, a t-shirt, and slicked-back hair showcased his sharp 1980s style but also quietly revealed his profession. It was 1987 and Hamid had just left war-torn Iran. Hoping to emigrate to the United States, he landed in Ankara, Turkey, where he had arranged for a visa interview at the American Embassy. As they dug through the details of his personal life, officials asked Hamid about his line of work. "I am an underground video dealer," he said cautiously. Worried his candor would ruin his chances of getting a visa but also afraid to lie, he braced himself for the worst. Instead, the agents erupted into laughter. Hamid's face must have signaled his confusion, because one of the agents finally managed to say, "It's just a silly profession, is all." But for Hamid, his work was far from a laughing matter. Dealing in home video technology of any sort was illegal in Iran at the time and could land him in jail. Hamid had nevertheless risked his freedom to establish a small underground video rental business. From a working-class background and only in his late twenties, he made a good living for himself and even provided for his family.

Thanks to a friendly face, he passed through the streets of Tehran unnoticed with a large briefcase full of Betamax cassettes in tow. His short list of clients, who only knew him as Hamid the Filmi, relied on him to stop by once a week with a new selection of movies on video for rent. His collection included everything from old Iranian films to Indian musicals to the latest Hollywood blockbusters. His clients would devour up to ten movies a week and then anxiously call him to bring more. As the war with Iraq raged on and Tehran became a target for bombings and missile attacks, these movies were a lifeline of sorts. As much as Hamid's business was profitable, he also thought he was doing important work. He didn't just deal contraband items on the black market; he delivered entertainment and escape and distraction at a time when people needed them the most.

Hamid eventually made it to the United States, where he would regale his friends with stories of the embassy officers' laughter when they heard what he did for a living. But something about their laughter never quite sat right with him. It was as though the Americans did not understand or take seriously the stakes of his work: the personal risks that he took and the importance of videocassettes to his clients. Perhaps those American agents took the availability of videocassettes for granted. The legal disputes between technology companies and media conglomerates that were taking place in the United States in the mid-1980s certainly did little to affect their access to movies on videocassettes. But the story of video is not the same in every country, and entire generations of Iranian media consumers came of age through the underground networks of which Hamid's small business was a part.

Hamid's uncomfortable interview at the American Embassy in Ankara suggests how easily people misunderstood the work involved in distributing videocassettes underground in Iran. That incident did not just result from the American officers' failure to acknowledge the precariousness of Hamid's work. For his part, Hamid was anxious about revealing what he did because the state had categorized his work as criminal. Such an encounter pivoted on the different meanings that people projected onto the work of video distribution. For the Americans, dealing in videos was frivolous work but worthy of a good laugh. For the Iranian state, it was an obscene crime whose punishment might involve prison time or even severe lashings. But for Hamid and the thousands of other Iranian video dealers at the time, video distribution represented something entirely different. Through their work, video dealers got to be part of the movies and to have valuable opinions about art and culture. They also forged a rare path to financial success during unprecedented economic hardships. In short, video dealing was a form of aspirational work.[1]

The work of video distribution in this period constituted a unique form of media labor, one that bears further theorization. Daniel Herbert has shown how, in the heyday of the American video store, the video clerk served as an arbiter of taste as much as a customer service representative.[2] In Iran, too, the labor associated with the informal video industry played an important role in mediating people's relationships to videocassettes and to movie culture as a whole. Video dealers bore the responsibility of transporting contraband items between private and through public spaces. Beyond this crucial act, they also served as tastemakers, curating the content that people were able to consume and instructing them on how to interact with a new media technology. It would be easy to disregard video dealers as intermediaries responsible only for the logistics of moving videocassettes from one place to another. In this chapter, however, I draw on oral history interviews, feature films, documentaries, essays, and trade publications to demonstrate that video dealers were actually cultural laborers. I show that in order to understand the work they performed, we need to foreground the creativity that marked and motivated almost everything they did.

This form of cultural labor was facilitated by and necessary to the underground video infrastructure. As I've shown in previous chapters, the video ban made possible new ways of belonging in Iran. The ban may have foreclosed certain possibilities but it opened up others. The informal video infrastructure was the site of these new subjectivities. When video distribution became a crime, most legal video stores abandoned their operations. Many of the people involved in the formal video industry of the early 1980s left the business of videocassettes altogether in 1983. This created an opportunity for an entirely different kind of person to be involved in the cultural work of dealing in videos, especially with the growing demand for movies on videocassette in the country. Working-class Iranians, in particular, stepped into the work of video distribution. This entry would have been impossible before the ban because they lacked the financial and cultural capital necessary to operate within the formal video industry. With the ban, however, a new class of video dealers accepted the risks associated with the job in order to perform aspirational creative labor that enhanced the meaning and value of videocassettes as they circulated underground. Ultimately, this book is about the vibrant lives that people build beneath repressive institutions, and video dealers—through their aspirational labor—played an important role in *making* culture for everyone who participated in the underground.

The cultural work of video distribution depended on the materiality of video. It was, after all, the bulkiness of movies on video that demanded the dealer's labor in the first place. Video dealers had to conceal the boxy cassettes while also knowing which routes to take to minimize their encounters with the police. The wholescale ban, which applied to all video technology rather than any specific content, demanded expertise in safely and secretly moving material goods. The material properties of video may have made dealing dangerous but they are also what allowed it to be creative work. This was the very contradiction of the ban at work. The underground could be both a site of precariousness but also one of astonishing creativity.

In this chapter, I map out what the work of video distribution meant to dealers and their customers and for media consumption more generally. Challenging the idea that the videocassettes circulating underground were merely copies, I locate the creativity implicit in the work that video dealers performed. That creativity was tied up in the precariousness of their work, and it allows for rich inquiry into what is allowed to be called cultural labor in the first place. Today's new economy—which treats creativity as a commodity—has transformed many forms of cultural work into invisible, unregulated labor.[3] A historical case like that of the Iranian video dealer attunes us to the many entanglements of creativity, labor, and subjectivity.

A CRIMINAL ENTERPRISE

The labor of underground video distribution did not just appear out of thin air when the Ministry of Culture and Islamic Guidance (MCIG) issued its ban on home video technology. It grew from the rapidly expanding video industry of the early 1980s and was, in some respects, a continuation of a long history of informal media distribution in the country.[4] In order to analyze this form of media labor, we need to understand how video dealing went underground. The question is not how the ban created video labor but rather how it transformed this labor. What did video distribution look like before the ban and how did it become a criminal enterprise under the ban?

Between 1980 and 1983, the labor of video distribution was marked by quick growth and increasing formality. This brief period saw the rise of trade publications, syndicates, and technical expertise, but also of regulatory policies that sought to control the labor involved in the budding video industry. When the ban struck in 1983, it redirected the energy of video distribution. Suddenly, video dealing became completely informal work marked by its mobility as much as its criminality. In this new system, individual agents illegally transported movies on video to customers. If video dealing before the ban was encapsulated in the traditional brick-and-mortar corner store, then video dealing during and after the ban was a diffuse, unregulated network of mobile laborers.

Between 1980 and 1983, during the height of the legal video industry in Iran, video dealing became a distinct form of labor. The conventions of the profession within this three-year period set a precedent from which underground video dealing would later emerge. Video distribution grew quickly and uncontrollably in the early 1980s, as small-business entrepreneurs invested in movies on video. Alongside this rapid growth, institutions and regulatory practices swiftly formalized the video industry. Like elsewhere in the world, the earliest video distribution efforts in Iran came attached to existing sites of consumerism, like hardware stores and corner grocers.[5] It wasn't long, however, before the popularity of videocassettes demanded stand-alone stores. The shift to a specialized marketplace signaled a transformation in the profession, as the distribution of videocassettes distinguished itself from the distribution of other kinds of commodities.

As specialized video stores proliferated in cities throughout the country, so too did the need to centralize their operations. This proved true especially from the government's perspective. As the number of stores grew, it became increasingly difficult for the Commission Against Social Evils (CASE)—the government institution charged with overseeing the video industry at the time—to control each individual store. As a result, it insisted on a syndicate to serve as an intermediary between the increasing number of laborers and the state's regulatory institutions. CASE temporarily closed all video stores for several days in 1981 until a governing organization had been created. In the end, the Association of Video Club Owners (Ettehādiyeh-ye kolubdārān-e vide'o) was founded by ten members on August 23, 1981, with Esmail Ahmadi as its first director.[6] Also known as the Guild for the Production, Distribution, and Duplication of Permitted Videocassettes (Senf-e towlid, towzi' va taksir-e navār-hā-ye majāz-e vide'o), the syndicate's establishment signaled that video dealing had become a distinct form of labor with its own challenges, obstacles, governing principles, and ethics.

The syndicate also marked the growing professional opportunities within the formal video industry. It sought to protect the interests of video club owners while also serving as a liaison between video club employees and the state. On the one hand, the union was an advocacy organization that sought to improve the quality of the profession. In particular, the organization took measures to enhance public perception of video technology, since many Iranians at the time remained skeptical of it.[7] On the other hand, the syndicate acted as a mechanism through which the state implemented its policies by insisting that all video clubs register with the organization. Within the first several months, 1,500 clubs registered with the syndicate. That number would ultimately rise to 2,000. At that point, all unregistered video clubs were shut down in accordance with state policy.[8] In addition to video store employees, scores of other professionals were joining the ranks of the budding video industry, including repair men and videographers.[9] The rise of video at this time meant not only an increase in access to movies but also new economic opportunities and professional ventures.

As video distribution became a legitimate form of work, modes of knowledge production emerged to service it. In contrast to the scant references to video in the major media trades of the late 1970s,[10] the early 1980s saw entire publications dedicated to video, including booklets such as *Vide'o che filmi bebinim?* (What should we watch on video?) and *Salām Vide'o* (Hello, video). These small runs included information about permitted movies on video and instructed employees about the videocassettes available for rent. These early publications show how video store clerks served as intermediaries between customers and movies. Video store employees taught people about cinema and home video technology, as much as they sold a product. These publications also document the trajectory of video dealing as it evolved into a more organized and formalized profession.

The formalization of the profession peaked in July 1982 when *Sinemā dar vide'o* (Cinema on video) published its first issue. Founded by Massoud Mehrabi, Houshang Golmakani, and Abbas Yari, *Cinema on Video* was intended as the trade publication for the Association of Video Club Owners. It published lists of permitted movies for video distribution, included technical information on videocassettes and video players, and featured interviews with high-ranking officials such as the head of CASE. This diverse content represented the various kinds of knowledge required of legal, formal video dealing—including cultural, technical, and regulatory expertise. At the same time, the title of the magazine clearly positioned the world of video squarely within the domain of cinema. The founders of *Cinema on Video*— all art school graduates and film critics—lent the profession of video dealing a degree of cultural cachet that was not necessarily inevitable. More than anything, however, *Cinema on Video* showed how video distribution was becoming a legitimate profession—one that operated at the nexus of cultural and technical expertise.

With organizations like the Association of Video Club Owners and publications like *Cinema on Video*, the formalization of video distribution supplied support, resources, and community for the growing number of employees who worked in the field of video. On the ground, the first generation of legal video dealers also felt the energy surrounding video during that period. For example, I spoke to Maani Petgar, a filmmaker based in Tehran, who worked in video distribution immediately before the ban. With a longstanding interest in movies, he opened a video store in 1983, just as video rental was starting to take off on a large scale. He had previously owned a photography store, but the new technology of video was too enticing to pass up. During our conversations, he likened video to a "miracle" (*mo'jezeh*), not just because of its technological affordance—like rewinding and rewatching—but also because of the role it played in a country that seemed increasingly isolated from the world. As he would tell me, "At the time, I often thought about how terrible it would have been if videocassettes hadn't existed. We would have been completely cut off from world cinema." Opening a video store was a chance to be part of video and the cultural and technological world it promised.

Owning a video store during this period required financial and cultural capital available only to a certain class of entrepreneurs. Formal policies and procedures structured the daily operations of Petgar's store. In order to open the store, he bought an existing video store charter from an acquaintance for a staggering 30,000 tomans. To put that price in context, Petgar explained that his monthly rent at the time was 4,000 tomans for a one-bedroom apartment with parking in northern Tehran. Once his store opened, he and his colleagues depended mostly on formal channels for their stock of videocassettes for rent. This was the most price-efficient option, even if the videos were censored and dubbed in Persian. It was too expensive to own a Sony Betamax T7 recorder, which would have allowed the store to prepare its own cassettes. Besides, video distribution companies at the time allowed stores to pay for the videocassettes *after* they recouped the cost from rentals. There was also demand for uncensored movies on video in their original language, so the store carried a few such movies that they rented under the table to customers they knew well. This wasn't much of a stretch for Petgar, who had been selling underground copies of his expansive rock music collection for years. Moreover, the formal structures of the video store business protected the under-the-table deals that took place there.

Although the video store was certainly a commercial enterprise, Petgar preferred to defer to his business partner on those matters. For Petgar, the video store was a cultural space, and he valued working with customers and helping them select movies to rent. He purchased video trade publications like *Video* and *Cinema on Video*, but he also had a large collection of film periodicals from before the revolution that had introduced him to arthouse films by the likes of Godard, Bergman, and Fellini. As he told me, "Considering my age, I was quite knowledgeable about world cinema, its directors, and its genres." The video store was a cultural space not just because of what it held but also because of who gathered between its walls. Petgar told me that Amir Naderi was one of the first customers to frequent the store. Naderi would go on to become one of the iconic filmmakers of this period, especially with the international success of his film *Davandeh* (The runner, 1984). Petgar would get his start in the film industry working on the set of *The Runner* as a still photographer and assistant director. This development shows how the video store was a site where important networks were forged and fostered among a certain cultural milieu.

When the ban on video technology was issued in 1983, all of the resources dedicated to video labor suddenly disappeared, at least officially. The syndicate collapsed, the support technicians went out of business, and *Cinema on Video* rebranded itself as *Film*, which ultimately became the longest-running film trade publication in the country. Overnight, video distribution had gone from a highly regulated profession to a criminal enterprise with an entirely unregulated workforce. This transformation played out in unexpected ways. In some respects, the most logical course of action would have been for the existing video clubs to continue their work underground. After all, they already had the knowledge, motivation, customers, and supplies of videocassettes. I had expected that the first generation of underground video dealers would be former video club owners. But the new precariousness of the work—its instability and the personal risks it required—deterred most people involved in the formal video industry from continuing their work once video was made illegal.

Petgar, for example, was forced to shut down his video store when the MCIG announced the ban on video technology. Still committed to the idea of providing access to movies on video, he began delivering his stock of videocassettes to former customers. As he told me, his income "skyrocketed" immediately after the ban because the demand was so high. In the end, however, the financial gain was not worth the hassle. Underground dealing was a far cry from the under-the-table dealings that had happened within the safe confines of the video store. According to Petgar, every time he would leave the house to deal videocassettes, he would first passionately hug and kiss his new wife like a soldier going off to war. The risks never went away. Petgar remembers his anxieties doubling when he learned that many video dealers had ended up in Evin—Iran's most notorious prison. His turn eventually came, too. A colleague ratted him out to the police, but Petgar wasn't at home when they came for him. Instead, they arrested and held his business partner until Petgar turned himself in.

In the end, Petgar and his partner managed to avoid jail time but it required a significant bribe. The police officer agreed to expunge their record, and in exchange they gave him a Sony Betamax T7. This exchange shows the currency that video technology held at the time, but also the irony of the underground. It would be easy to view the state and the underground infrastructure as diametrically opposed, but the reality encompassed more shades of gray. With his own Sony Betamax T7, the police officer that Petgar bribed was ready to participate in underground video culture himself. For Petgar, the scare was enough to deter him. He sold his videocassettes to someone more willing to take the risk and eventually emigrated to Australia, where he lived for ten years. Petgar's story shows how the nature of video distribution changed in 1983 with the implementation of the ban on video. The updated demands of the work encouraged a new kind of dealer with an entirely different profile and unique motivations.

Given the decentralized nature of video dealing at the time, it would be impossible to suggest a single profile for video dealers. That being said, during the ban period many—although not all—video dealers came from a working-class background and had day jobs in addition to their work in video distribution. Underground video dealers were moonlighters. They were free agents unbound from permits, bureaucratic procedures, and physical stores. Video distributors of the early 1980s had been highly visible with their brick-and-mortar stores, flashy signs, movie posters, intricate displays, and a growing number of publications dedicated to their profession. In contrast to these early forms of video labor, underground video dealing beginning in 1983 revolved around invisibility and the diffusion of its community and its resources. Existing stores quietly sold their stock of movies and video equipment to individuals willing to take the risk. With these sales, the old order broke down and gave way to new forms of video labor. The new underground video dealer took on an array of responsibilities involved in preparing and transporting movies on video—everything from duplicating to dubbing to delivering.

As video dealing shifted from a highly regulated profession to a criminal enterprise, the popular discourse surrounding it also changed. Between 1980 and 1983, owning a video store was emblematic of the rhetoric of modernization that had circulated in the country for decades. For early investors like Petgar, owning a video store married middle-class entrepreneurial aspirations with cultural capital. Around the world, marketers and enthusiasts likened home video technology to a revolution that would propel society forward.[11] In Iran, too, the earliest references to video in the late 1970s call it a revolutionary technology. Owning a video store was an opportunity to be a part of an innovative technology that promised to change the world. This was certainly the impression of the earliest video store customers, who viewed the profession as heroic.

Farhad was a teenager when the first video store opened in his neighborhood in central Tehran. As Farhad would tell me, the owner of the video store was considered a hero among the neighborhood's schoolboys. He was in his early thirties with a trademark thick mustache. He regaled his customers with his knowledge of movies but also spent hours explaining the inner workings of the video player to anyone who would listen. For Farhad and his friends, the video store owner embodied the qualities they wanted for themselves: modern and cultured. He told me, "The video store owner was the coolest guy we knew. We all tried to grow mustaches like his, but our adolescent attempts seemed feeble compared to his."

The rhetoric of cultural and technological advancement associated with early video distribution labor changed radically in 1983 with the ban. As the state demonized home video technology as an obscene agent of cultural corruption, the video dealer became synonymous with a life of crime. The associations between video dealing and crime never disappeared. As a result, video dealers remain the most misunderstood characters in the story of videocassettes, despite the central role they played. In both official sources and popular lore, a common motif in depictions of the video dealer was that of a mafia member. Official sources go so far as to transliterate the word "mafia" into the Persian script,[12] evoking a sense of organized crime that may have had very little to do with the historical reality. There is some evidence to suggest that the ban on video forced cassettes into existing circuits of contraband items, which may have been controlled by an urban mafia. If this was the case, however, it did not last long. Video distribution quickly became decentralized as individual dealers set up their own informal businesses with no affiliation and surprisingly little concern for competition.

The discourse on video distribution, nevertheless, hinged on the idea of professional crime, especially in popular sources like movies. The video dealer as career criminal became the dominant motif through which many Iranians understood this particular form of work, especially during the ban period. This was, no doubt, to the delight of the state, which sought at the time to link video to immorality and corruption to further its own agenda. Official sources thus frequently referred to video dealers as *qāchāqi*, a word that means both "smuggler" and "bootlegger" in Persian. A word like *qāchāqi*, which was largely used with regard to drugs, both criminalized the work of video distribution and emphasized its associations with the movement of other illegal material.

These characterizations of video distribution are consistent with popular representations at the time. The movie *Āvāz-e Tehran* (Song of Tehran, dir. Kamran Ghadakchian, 1991), for example, exposes Tehran's criminal underbelly. Rahman, a recent arrival from the countryside, immediately gets swept up in the illegal world of video dealing, where he encounters veteran criminal Fereidun. The film is replete with underground dealings and arrest scenes (figure 3.1a, b). Moreover, Fereidun's untimely death as a direct result of his crimes sends an explicit message at the end of the film about the kind of criminal life that video rental supports. *Song of Tehran* clearly links video dealing to the social ills of urban life.

All films in Iran go through a stringent process of censorship, so *Song of Tehran*'s stance on video distribution stays predictably in line with the state's overall rhetoric on video distribution. The movie is a rare instance in which the state acknowledged and even visualized the existence of an underground world of video dealership. Despite the film's occasionally sympathetic lens, it remains resolute in its condemnation of the work of the video dealer as a criminal figure.

Figure 3.1a, b
In *Song of Tehran*, (a) two men are arrested for their work dealing videocassettes, and (b) the main character Fereidun wields a knife in front of a stack of videocassettes as he prepares to kill the big boss. Scenes like these emphasize the criminality and immorality of video distribution.
Source: Āvāz-e Tehran (dir. Kamran Ghadakchian, 1991).

Oral history interviews and essays also confirm the ways in which the video dealer became inextricably tied to the criminality of his work. The video dealer's association with professional crime continues to mark how some people remember the work he once performed. For example, I asked one interlocutor if he knew of any famous video dealers. In response, he told me the story of Jamshid. According to this urban legend, Jamshid was a criminal boss who had made a small fortune specializing in products banned by the Islamic Republic, including alcohol, music, and drugs. When video-cassettes and players were banned in 1983, he poured that small fortune into the informal video market. And it required a small fortune because the ban had temporarily inflated the cost of all video equipment. Jamshid, who, as my interlocutor adamantly emphasized, was not even interested in cinema, quickly came to control the video market, not just the equipment like cassettes and players but a union of dealers, as well. His investment paid off; he multiplied his original fortune many times over in just a few years. Satisfied with his earnings and bearing no real attachment to video technology, Jamshid abruptly left the market and retreated to a luxurious seaside mansion, where he continues to live comfortably on the money he made during his brief foray into the video rental.

I refer to the story of Jamshid as an urban legend because the ways in which it has been told and remembered are more important than the truth of its details. After hearing this story for the first time, I began asking other interlocutors if they knew of Jamshid. Many had, although no one knew him personally. My many unsuccessful attempts to locate Jamshid made me question whether this legend was based on a real figure or rather represented a compilation of people's anxieties about the video dealer. While the details of the story shifted, as Jamshid became Javid and a seaside mansion became a villa in the Swiss Alps, the core of the story remained the same. It underscores popular perceptions of video distribution in the 1980s, at least as they remain in an Iranian collective memory. The video dealer was a hardened, career criminal, whose motivations were purely financial, and the structure of video distribution was organized but corrupt. The payoffs were high, too, according to this urban legend. They were enough for a comfortable pension—at least for those at the top.

Forms of collective remembering, like this urban legend, affirm how the work of video distribution has been memorialized in such a way that it remains indistinguishable from its early criminal life. While many of the everyday users with whom I spoke were resolute in their belief that watching videocassettes was not immoral or criminal despite the ban, they were also steadfast in their categorization of video distribution as inherently criminal. While Iranian society detached the consumption of videocassettes from the state policies that banned them, attempts to salvage video's cultural value often left the video dealer in the dust.

More recent written accounts also link the video dealer to his crimes. For example, in her piece "The Video People," essayist Media Farzin tries to remember the people behind the scenes of video distribution.[13] She describes the importance of videocassettes to the generation of Iranians who grew up in the 1980s and 1990s, but most importantly she foregrounds video dealers as central to how everyday Iranians accessed the world of videocassettes. As she humanizes the video dealer as someone who would sometimes stay for tea, she necessarily falls back on the language of the black market. She writes, "There were stereotypes of black market men, shifty-eyed opportunistic types you wouldn't want to meet alone in a back alley, especially if you were a girl."[14] She goes on to explain that video dealers didn't conform to this stereotype at all. Elsewhere in the essay, however, she refers to video dealers as "black-market dealers in moving images" and "black marketeers."[15] Far from criticizing Farzin's thoughtful piece, I want to suggest that in order to redeem the figure of the video dealer, she has to evoke his criminality because this is how most people understand him. This language, and Farzin's essay as a whole, reminds us how deeply the associations run between the video dealer and the black market.

On the surface, it would appear that the labor of video dealing transformed significantly with the ban, and to a certain extent it did. Video distribution became a completely underground pursuit: unregulated and criminal. In practical terms, this meant that underground video dealers had no formal, legal framework through which to operate their small businesses. The sense of community and organization that had pervaded the profession between 1980 and 1983 no longer existed. With these transformations, criminality also became the dominant motif through which ordinary people understood the profession. Popular discourse at the time, however, largely missed the mark when it came to describing the mechanics of underground video dealing. While the illegality of video distribution made the work risky, I will show that video dealing represented much more than the pursuit of a career criminal. Just as before the ban, underground video dealing was aspirational work whose operations teetered between business savvy and cultural knowhow and crossed formal and informal practices and spaces.

AN OPPORTUNITY FOR ACCESS

The case of the video dealer in Iran stands to teach us a great deal about the conditions driving informal labor. As more and more labor shifts into informal economies, we need rigorous accounts that describe a full range of experiences therein. A revolution, then a war, and finally a wholescale ban on home video technology created an environment in which everyday men and women risked their freedom and personal safety in order to move videocassettes through an underground market and into people's homes. It would be easy to understand this informal labor through its criminality and its precarity. As we just saw, that was the dominant way of configuring underground video labor. But the actual experiences of video dealers ask for something else altogether. While the informal nature of their labor did create precarious work conditions, this precariousness was as much an opportunity as a limitation. Video dealers performed this work because it was tied up in their identity and their aspirations. Far from being the victims of an unfair system, they were knowledgeable of the risks they were taking and strategic about the opportunities the informal market offered them. Underground video work offered them access to social and cultural worlds that would not have been available to them if the video industry had continued to formalize throughout the mid-1980s. Video dealers were distinct from other kinds of dealers (e.g., drug or alcohol), because as much as videocassettes were material objects that had to be transported, they were also cultural objects that needed to be mediated to consumers.

To a certain degree, the aspirations of underground video dealers were also political, as video dealers disrupted the social and cultural structures that had supported successive ruling classes both before and after the revolution. In this respect, the underground video dealers of Iran shared certain qualities with the worker-poets who animate Jacques Rancière's *Nights of Labor: The Workers' Dream in Nineteenth-Century France* (1981). Rancière studies working-class men and women in the 1830s and 1840s, who labored during the day and wrote poetry and philosophy at night.[16] He gives meaning to their creative ambitions by pushing back against the idea that workers in the mid-to-late nineteenth century simply wanted better working conditions. As he shows, they wanted different lives altogether. More than anything, they desired the luxury of time, which had been stolen from them by the dominant idea that "good workers" labor during the day and sleep at night to prepare their bodies for more labor in the morning. By taking back their nights, these men and women not only voiced their concerns through their intellectual work but also disrupted the very structures of time that had limited their lives in the first place.[17]

Just as the working class had been at the center of the French Revolution a century before the worker-poets in Rancière's study, so too had the working class been central to the Iranian Revolution less than a decade before video dealers began undertaking their work. The revolution in Iran, however, did very little to improve the lives of working-class citizens. On the contrary, a weakened economy meant a higher than ever unemployment rate, while the war with neighboring Iraq led army recruiters to target working-class families in order to send their men and boys to the warfront. While the revolution succeeded in overturning the monarchy, the theocratic structure of the new Islamic Republic simply replaced one ruling elite with another, leaving most working-class people behind. It is within this context and in light of Rancière's discussion of worker-poets in nineteenth-century France that the work of video distribution takes on new meaning. The video dealers in Iran did not just take advantage of the Islamic Republic's rigid policies for more income, they also subverted them in order to become creative workers themselves, to operate outside of a rigid class structure that had determined their lives both before and after the revolution.

This rigid class structure had been reflected in movie culture for decades in Iran. As early as 1936—just as movie-going culture was developing in Iran as a widespread public phenomenon—the state codified the experience of watching a movie as a class-based experience. Through its "Regulations for Cinema," the state designated three classes of movie theaters. First-class cinemas were to start with nonfictional programming, including documentaries and newsreels, and would usually feature foreign films, often from the United States and Europe in their original language. Meanwhile, second- and third-class cinemas showed popular films, including local Iranian productions and foreign films from countries like Egypt and India dubbed in Persian.[18] Poorer areas in Iran had few movie theaters in the 1930s and 1940s, and they

tended to carry second- and third-class designations.[19] This
grew more rigid with time. By the mid-1970s, most second- an
cinemas could be found in working-class neighborhoods.[20] Th
system for movie theaters shows how viewers' socioeconomic back
determined the kinds of movies that they could watch.[21]

Class did not just determine who had permission to watch certain kin
of films but also who could participate in a film's production. This is evident
in the 1978 film *Sorkhpust-hā* (Indians, dir. Gholamhossein Lotfi), which
chronicles the cinematic dreams of Tehran's working class.[22] Released a year
before the revolution, it tells the story of a group of working-class men who
aspire to be in the movies. Every day they gather in a coffee shop on Arbab
Jamshid Street in Tehran's Lalehzar district, a film production hub at the
time. There, they wait for a man called Chichu to come and bring them to
the set as extras. As they wait in the coffee shop, they imagine entirely new
lives for themselves—much bigger than their current social class allows.
They fantasize about rubbing elbows with Behrouz Vossoughi—the biggest movie star at the time—or acting in a scene with the esteemed Reza
Beyk Imanverdi. With names like Ahmad the Actor and Mostafa James
Bond, their dreams of cinema have consumed their entire identities, even
their names. As the movie progresses and they continue to wait day in and
day out, their dreams get smaller and smaller. At this point, they would be
happy with just taking a picture with Beyk Imanverdi. In the end, however,
no one comes for the men and they never make it into the movies. A film
like *Indians* documents how the Iranian film industry historically shut out
entire classes of people. As much as popular cinema in Iran was targeted to
working-class viewers, they were not allowed to participate in its production. They could only be passive viewers of certain kinds of films.

Many of these same tensions remained a decade later, even after a revolution and war had restructured the film industry. Indeed, the cinematic
desires of Iran's working class lie at the center of Abbas Kiarostami's well-
known film *Close-Up* (1990). Blurring fact and fiction, the film chronicles
a real-life legal dispute between Hossein Sabzian and the Ahankhah family.
After a chance encounter on a bus, Sabzian convinces the family that he is
the renowned filmmaker Mohsen Makhmalbaf. He even tells the family
that he would like to feature them in his next film. This promise aligns perfectly with the cinematic ambitions of the family's son, Mehrdad, who "has
chosen art instead of selling bread" like his older brother. As an aspiring

...ak. In pretending to be Makhmalbaf, Sab-

...hrdad to play out their fantasies of enter-

...sies come to an end, however, when the

...stencies in Sabzian's story and have him

...and at the judge's recommendation, the

...inst Sabzian. Forgiving him in a speech,

...nes as the result of his unemployment.

...h Kiarostami, however, Sabzian claims

...different. "I am not a crook," he insists. When Kiar-

...stami presses him to explain what he actually is, Sabzian resigns, "I am just really interested in cinema."

Close-Up constructs a powerful message about how the Iranian film industry is set up to exclude certain classes of people. Sabzian is an unemployed father, and Mehrdad comes from a family of bread sellers. Their ambitions are much bigger than their socioeconomic status allows for. In a move typical of Kiarostami, both men end up finding their place in the cinema not only as subjects of the film but also as actors who reenact the events of their own story. Following the success of *Close-Up*, Sabzian experienced his own fifteen minutes of fame but ultimately slipped back into obscurity and poverty. He spent the last years of his life as a video dealer at a bus station in south Tehran. In the 2005 documentary *The Film School of Hossein Sabzian* (dir. Azadeh Akhlaghi), he admits, "I dreamt of being a general but ended up a foot soldier." Sabzian's life both within *Close-Up* and beyond shows how underground distribution offered opportunities to work adjacent to the movies, especially to those who were otherwise shut out from the film industry.

Indeed, the ban on video technology—intended as a restrictive policy—actually opened up space for social mobility by placing control over the distribution of movies in the hands of working-class men and women. I spoke with Hussein, an early underground distributor of videocassettes, who embodied the social-creative aspirations of dealers. In 1984 he bought a stock of videocassettes from an acquaintance who had decided to leave the country. In his mid-twenties and freshly returned from the war front, he used the cassettes to set up a small business renting out movies. For several years, he labored during the day as a lineman for the government-owned telecommunications company, and distributed videocassettes after work. The telecommunications company gave him a van to transport the

113

equipment needed to service telephone lines, an important job since the war meant that the city's infrastructure was constantly under attack. The van also proved to be the perfect cover for his work moonlighting as a video dealer. It provided access to the entire city and immunity from the police, who never searched service vehicles like his.

When I spoke to Hussein in 2016, two things very quickly became apparent. First, he was exceptionally knowledgeable about cinema, although my presence as someone who had identified as a media scholar made him self-conscious of that knowledge. Second, his work in video distribution was not motivated by money. This latter point threw a wrench in my own hypothesis that the early generation of working-class dealers were primarily driven by financial reward. Instead, I came to realize that Hussein had at stake something much more valuable than money. Through his work as a video dealer, he took advantage of the cultural capital associated with films and aspired to a kind of social mobility that would have otherwise been closed off to him.

At the time, Hussein lived in south Tehran, which comprised predominantly working-class neighborhoods. Most of his clients, however, lived in the northern and western parts of the city, which were typically middle and upper-middle class. As in most places, the early adopters of videocassettes in Iran were the affluent, those with the resources necessary to invest in a technology whose novelty and scarcity drove the market value up. Through his work as a video dealer, Hussein would literally move up, as he drove from the south of the city to the north in order to distribute videocassettes. But this movement wasn't merely physical. It was also social. Once inside a customer's home, he was no longer just the person who delivered videocassettes; he was a film expert. His clients depended on him for suggestions and critiques. His knowledge mattered in their world and it was valuable. As a lineman Hussein was of course very knowledgeable about telephone systems as well, but that knowledge bore association with unskilled labor, whereas knowledge of art and culture was a luxury only afforded by affluent life. For several hours each day, Hussein became part of a new social class, performing his knowledge of cinema in a way that proved meaningful to those around him.

Even when he returned home to south Tehran, his life embodied the social mobility to which he aspired. He spent most nights with his friends at cafes, where he would treat his entire group of friends with the money he had earned dealing videocassettes. While Hussein usually failed to impress his friends with his commentaries on film, he nevertheless remained at the helm of his social network because he bought them the luxury of time. Just as the worker-poets in Rancière's study weren't seeking merely better working conditions but rather entirely new lives, Hussein didn't want a better material life but rather the luxury of time and the space to engage in conversations about culture—two features that he associated with middle-class life.

As video dealers in Iran entered the informal economy, they started to conflate their work with their social and cultural identity. This becomes apparent when we compare Hussein's experiences to those of an entirely different kind of video dealer. On the surface, Siamak appeared to be everything Hussein was not. He came from an upper-middle-class family and earned a degree in engineering from the University of Tehran, one of Iran's most prestigious universities. Unlike Hussein, who counted among the first generation of underground video dealers, Siamak did not get into the business of videocassette distribution until the late 1990s—just as video compact discs (VCDs) were starting to overtake the analog videocassette market. Unlike Hussein, who dealt primarily in the popular hits of the day, Siamak specialized exclusively in arthouse cinema, everything from the classics to the obscure. Despite these differences, Hussein and Siamak's profiles are not as disparate as they might initially appear. As I will show, the work of distributing videocassettes was aspirational for both men: it marked the performance of an entirely new social or cultural identity.

Siamak began dealing videocassettes in the late 1990s at the urging of a friend, who thought his obsession with cinema would suit him for this work. Although the ban had been lifted by then, most titles were still not legally available on video and, thus, the underground infrastructure and its laborers continued to play an important role. An engineer by training, Siamak quickly established himself as one of the most prominent video dealers in Tehran, especially for those people closest to the film industry. He became the go-to video dealer for *Film* magazine, the most important film trade in Iran. In his personal dealings, he would also claim to work with big-name filmmakers. As one of his former customers told me, Siamak would often position his suggestions by referencing his most famous clients. He would say, for example, "Abbas Kiarostami recently rented this film and really liked it." Of course, at that time in the early 2000s, Kiarostami had recently won the Palme d'Or at the Cannes Film Festival, making him one of the most important filmmakers not just in Iran but in the entire world. This kind of posturing, for which Siamak became well known, speaks directly to his cultural aspirations. Through interviews with both Siamak and his customers, I came to appreciate his work not only as lucrative—although it was—but also as an attempt to break into a cultural world that otherwise kept him out.

For Siamak, nothing was more important than maintaining his identity as a dealer of high art. For example, one of his customers, Navid, told me that at one point his friend was looking for animated pornography. Lacking any leads, the friend asked Navid for suggestions. He took the request to Siamak, who returned a few days later with a videocassette that contained a midcentury Italian erotic thriller. For Navid, the video was Siamak's way of setting the record straight. Siamak was not the kind of filmi who worked in something as frivolous as animated porn; he dealt only in serious art.

Navid also had occasion to socialize with Siamak a number of times, and he told me that Siamak often spoke of sex and prostitutes. This additional context gave special meaning to Siamak's refusal to give Navid's friend pornography. Through his work, Siamak forged an identity entirely separate from his lived reality. In other interviews, I heard him described as a "homely" figure who could never have made a career in front of the camera. His customers understood his work in videocassettes as an attempt to forge a reputation as someone who was cultural and refined.

Siamak's relationship to videocassettes shifted in 2003. That year the state cracked down on the country's cultural producers and arrested many people close to the film industry, including journalists, film critics, and even video dealers. Siamak was among the people arrested. After he was released, he decided not to deal in videocassettes anymore. Not only were bulky videocassettes an occupational hazard but the technology was quickly falling out of favor. He later tried VCDs, DVDs, and Blu-ray, before leaving the business altogether. Yet Navid told me that when he had recently run into Siamak, he was still trying to offer Navid rare movies in an effort to impress him. The act of dealing in illegal and obscure movies had become such an important part of Siamak's identity that he could not abandon it altogether.

The film *Sāqi* (Dealer, dir. Mohammad Reza Alami, 2001) also relates to the creative aspirations of video dealers. Released in the final years of the videocassette's popularity in Iran, the film opens with a voiceover that declares, "My name is Reza. Reza Tehrani. I am one of the young people in this city, and like most of them, I had many dreams. I was in love with films, and I wanted to change the world through filmmaking. That's why I picked a related profession. I hoped that someday I would get lucky and meet someone so that I, too, could enter the world of cinema." At the precise moment that Reza mentions his profession, we see images of him gathering piles of videocassettes and shoving them into a soft briefcase. He is a video dealer who hopes his proximity to films will someday translate into a career as a filmmaker.

Throughout *Dealer*, Reza has clearly achieved some degree of financial success. As viewers, we see him flying through the streets of Tehran in his flashy orange Jeep Wrangler. Yet there seems to be something else teeming under the surface of his work in video distribution. His customers regularly remind him of the motivations that got him into this line of work in the first place. One asks, "Have you made your own film yet? When do I get to go to the movie theater to see one of yours?" These aspirations haunt Reza and his work, and they become his vulnerability. One of his customers, a dentist who is never satisfied with the movies that Reza delivers, convinces him that he will need to finance his own work if he wants to direct films. While Reza makes a comfortable living as a video dealer, the kind of success he craves would require a more lucrative position. The dentist thus encourages Reza to deal in drugs, which ultimately becomes his unmaking.

Unlike *Song of Tehran* a decade earlier, *Dealer* offered a nuanced portrait of the video dealer as someone with real motivations for entering this line of work. It wasn't just happenstance or a lack of other options, nor did the profession itself necessarily need to be criminal. The representations of video distribution are friendly exchanges that likely resonated with viewers at the time. As a state-sponsored film, however, *Dealer* still had its own axe to grind. It gave into and perpetuated two of the most common misconceptions about the video dealer. It both tied video distribution to financial wellbeing and viewed this kind of work as a gateway to more serious crimes like smuggling drugs.

Despite the ultimate direction the film takes as a commentary on the drug epidemic in the country, early scenes showing Reza's work as a video dealer attest to the ways in which the work was creative—or at least aspired to be so. In his dealings, Reza isn't simply a courier. He views his role as an important one. He not only provides access to movies on video but also helps his customers cultivate tastes and opinions. In one instance, Reza arrives at an apartment where a handful of young women greet him informally and dote on him. They're genuinely happy to see him. When he sits down with three of them and asks about last week's movies, they offer conflicting viewpoints, as they hold and caress the videocassettes in front of them. Reza responds to these comments by saying, "You know, I still don't understand what kind of movies you guys like." This comment sets off a flurry of animated gestures and competing opinions as the women discuss their favorite films, before the scene fades out and Reza races to his next appointment. In this scene and others like it, we see Reza trying to appeal to his customers' tastes. He always sits with them, sometimes drinks tea, and together they talk movies. These interactions make clear that Reza not only cares for his work but also exercises his own knowledge of and interest in cinema through his exchanges with his customers.

When I first started studying video distribution, the available sources—especially trades and oral history interviews—led me to understand the profiles of video dealers through a lens of financial gain. In such sources, the ban on all video technology made the work of video distribution risky, and the promise of money seemed like the only thing that could logically mitigate that risk. Indeed, sources from the time often emphasize the monetary value of the informal video market during the ban period, foregrounding the potential for making money. For example, an article published in *Sorush* in August 1993 titled "The Tale of Video in Iran" juxtaposed financial gain and the illegality of video distribution. The article reports, "Video activities were very difficult, and the people who did this work often faced the risk of being arrested by security forces."[23] Later it argues, "The people who undertook this work were not against culture [*zedd-e farhangi*] but were rather seeking good income."[24] Such a statement is in sync with other sources that align the dangerous work of dealing videos with high financial gain. That kind of juxtaposition structures a logic of video distribution that understands money as the only reasonable motivation for breaking the law.

This narrative of video distribution is enticing. It paints a seductive picture of working-class men risking their personal safety to deliver millions of videocassettes to users around the country. While this narrative initially captivated me, as I spoke to more and more former video dealers, I came to understand that such an analysis of video dealing reinforced my own misconceptions about class and the very nature of video distribution. Indeed, the *Sorush* article, although sympathetic throughout, still underscores the criminality of video distribution. Such a reading suggests that only people desperate for money would break the law to undertake this kind of work. It links poor values to poor people. While most of the dealers with whom I spoke verified that they made good money dealing videos, they often punctuated their statements by emphasizing that the money mattered very little to them.

For example, we might think back to Hussein, who told me that while he made good money, he never saved any of it. In those days, he was in his twenties and the economy was admittedly bad. He and his friends spent every night at cafes, and they had a rule that whoever happened to have a job at that moment picked up the tab for the whole group. His steady work as a video dealer meant that he usually treated his friends. Significant to Hussein's story—and to that of many other video dealers—is that he had a day job and only moonlighted as a video dealer. Although working class, he was not strapped for cash. While his day job took care of his basic needs, it was Hussein's work in video distribution that supported his active social life. A case like Hussein's challenges us as scholars to come up with more sophisticated models to understand what motivated video dealers, especially during the ban period.

Video dealers' aspirations could be found in their commitment to the profession, even with its many risks. All of the dealers with whom I spoke were arrested at least once, and they told me countless stories of other dealers who had been arrested. Some managed to escape jail time, lashings, or hefty fines; others were traumatized by their experiences in prison. Perhaps surprisingly, many of them stayed in the business of video dealing even after they were arrested. This persistence shows that the work of video distribution mattered to the people involved in it. For example, a video dealer named Alireza explained to me that he continued to rent out videocassettes even after being arrested. When I asked why, he sighed, frustrated that I didn't already know the answer. Finally, he exclaimed, "What could I do? Movies were my entire life. Sure, we got arrested sometimes. But we were also free to live the life we wanted, and not many people could say that."

For many video dealers, the creative labor of video distribution became part of their identities; it became integral to who they are. There is an important lesson here about labor and subjectivity. In the same way that postwar middle-class workers in the United States viewed their work not just in terms of stability of income but also as "social belonging, a public identity, a sense of well-being, and future aspirations," work in the informal economy of Iran also became, for many, tied up in their subjectivity.[25] In Iran in the 1980s and 1990s, many video dealers pursued their work in distribution because it provided them with something that their own class status and day jobs could not. Specifically, it allowed for a public-facing, creative identity that would have otherwise remained out of reach.

TASTEMAKERS, CURATORS, AND CREATIVE LABORERS

Underground video distribution in Iran forces us to reimagine our understanding of creative work, especially when it comes to media. The idea of creative work inspires images of artisans and bohemians living luxurious lifestyles. In popular imagination, creative workers occupy a vastly different space compared to the quintessential industrial worker in terms of their tastes, class associations, and work conditions.[26] Under these terms, it would be easy to dismiss the idea of video dealers as creative workers. Not only did their class associations and work conditions align them with industrial workers, but the work they performed also did not seem especially creative, at least at first glance. Yet video dealers were more than neutral intermediaries or couriers simply transporting material objects. The work they did creating, curating, and circulating videocassettes imbued these commodities with meaning and value. Video dealing required ingenuity as much as it demanded technical and cultural knowledge.

The notion of creativity has been central to how scholars study media labor as well as work within the creative economy more generally. Yet efforts to theorize creative labor have largely been incompatible with discourses on piracy, which position informal media work as the antithesis of creativity. Like the Iranian state in the 1980s and 1990s, today's intellectual property rights regimes view informal media distributors as crooks—the moral enemies of art and culture writ large. In contrast, I argue that video dealers exercised a creativity that helped shape the cultural value of the products they distributed. We can trace creativity not only in the work that video dealers did transporting videocassettes but also in the way they curated and prepared their selections. Such a reading of the work of video distribution blurs clear distinctions between what we consider the original and what we consider the copy.

Although it's true that video dealers distributed movies outside of formal media industries, the videocassettes they circulated were not just copies.

The quality was often highly disparate, but so too was the content. As I show, video dealers played a major role in how a movie was received, embedding it with values and meanings wholly separate from its original production. Both videocassettes and their circulation were the video dealer's own unique creations.

This argument builds off of a growing body of scholarship that seeks to uncover creative labor in unexpected places and in unusual shapes and forms.[27] Despite the diversity of this scholarship, it shares an unbending focus on formal media industries, in which the *production* of culture—whether a movie, a television show, or a video game—remains the most important objective. One way to extend this line of scholarship would be to propose that illegal video distribution represents creativity within an informal industry, in contrast to scholarship on formal media production. Such a proposition, however, may fall short of adequately accounting for the labor of informal media distribution. As I will show, the work of video dealers was much more complicated, challenging easy distinctions between informal and formal industries and between production and distribution. Examining creativity within video distribution expands the limits of what can be considered cultural or creative labor in the first place.

Creativity may be most evident in the formal production of media content through the efforts of directors, cinematographers, actors, and screenwriters. But cultural intermediaries—to borrow Pierre Bourdieu's term—also exert creative agency through the work of "providing symbolic goods and services."[28] In other words, informal media distribution might also be considered creative labor, even if it remains many times removed from the mechanics of formal media culture. As Keith Negus argues, we need more rigorous accounts of those people "who might be involved in 'articulating' production with consumption."[29] Moving away from Bourdieu's claim that cultural intermediaries form "a new petite bourgeoisie,"[30] Negus includes a range of laborers who work in the "enduring space between *production* and *consumption*," including studio executives and factory workers.[31] In order to globalize our accounts of cultural intermediaries, we might also consider informal infrastructures alongside the formal industries that Bourdieu and Negus study.

Indeed, given that *globally* most people access movies through informal means, we need to ask—as Ramon Lobato has—how we might move beyond discussions on the ethics of piracy in our scholarship.[32] Because informal distribution practices are especially prevalent in the Global South, to disregard them merely as unethical would be to discount entire populations around the world and render them immoral. As a way of moving the conversation forward, Lobato proposes different frames for understanding piracy—everything from resistance to access to free speech.[33] Within this model, he offers authorship as one entry point into thinking about informal distribution.[34] Lobato suggests that piracy might even be considered a form of authorship, given that all forms of media distribution inflect the meaning of cultural products. He argues that movies are "subject to an array of value-adding processes on . . . [their] voyage from studio to consumer."[35] In other words, the means by which a movie reaches consumers matters when we think about how they ultimately value it. This understanding of piracy intersects with other attempts in media studies to broaden the definition of authorship to encompass all kinds of "creative agency."[36]

Video dealers exercised ingenuity in subverting the law. They took advantage of the rigid legal system in Iran in order to create a space for themselves as cultural workers—a space that, as I have shown, was otherwise shut off to them. Dealers have told me that as much as a fear of getting caught shaped their work, a certain pleasure could be found in avoiding detection most of the time. They went to great—indeed, creative—lengths to ensure that they evaded the police. While an aluminum Samsonite briefcase was the most common way for these dealers to move cassettes, my interlocutors told me with admiration about instances in which video dealers used creative means such as laundry bundles, overcoats, and even wheelbarrows to keep the videocassettes invisible as they shuffled them through the streets. The effect of all of these maneuvers was the special appreciation consumers had for cassettes and the movies recorded on them.

The scarcity and illegality of videocassettes gave them certain financial value, but the risks that the video dealer skillfully mitigated imbued movies with additional cultural value, as well. In other words, just by moving videocassettes and providing people with access that they otherwise would not have enjoyed, video dealers gave movies meaning. For example, Media Farzin reports that even though she was a student at Tehran's well-respected Tehran University of Art, the school's library didn't have any films. For that, she writes, "we had to turn to our people."[37] On one occasion, hoping to learn more about "non-diegetic sound," she requested Margaret Duras's

India Song (1975). Her video dealer did not have any of Duras's films but recommended Alain Resnais's *Last Year in Marienbad* (1961) instead. Farzin felt honored "to be entrusted with" a recommendation from her dealer, even though, as she claims, she "had no idea the logic behind it."[38] The recommendation turned out to be a hit:

> I sat down to watch the film expectantly that night. It was everything I could have hoped for, even as it made no sense. . . . I watched it again, alone. Then I threw a film night—chips and cucumber yogurt, bootleg alcohol, takeout pizza, and way too much giggling, for my taste. I still had no idea what it was about. . . . "I loved it," I told J [the video dealer] the following week.[39]

The way in which Farzin structures this narrative underscores the meanings and values that the video dealer's work lent movies. Beginning with a lack of access in her university's library, she then unpacks how the video dealer's presence and his intervention in her movie selection ultimately led to a deep engagement. Even though she never really understood the film or why her video dealer had recommended it to her, she valued it precisely because of how she had gotten hold of it.

In Iran in the 1980s and 1990s, information about cinema was scarce. This was before the arrival of the internet in the late 1990s. As a result, very few means of accessing information about cinema existed. University libraries occasionally carried film journals like *Sight and Sound*, and local publications like *Film* and *Film Report* did their best to keep film enthusiasts abreast of both the latest releases and the classics of world cinema. At the same time, there were no centralized databases or specialty video stores that could help average users access specialized information about world cinema. Video dealers filled that lacuna. Just as Daniel Herbert explains that the American video store trained Americans as consumers of media,[40] the video dealer helped instruct Iranians on what it meant to select media. While they were still limited to a video dealer's particular collection, the ability to choose or select what media they wanted to consume was an exciting new phenomenon.

Many of my interlocutors explained their passion for videocassettes through their frustrations with the available programming on state-run television channels. In the 1980s, the war effort took priority on the country's two main television channels, where programmers devoted most of the day to reporting on the ongoing war with Iraq. Images of guns and violence and statistics about casualties were commonplace on the television screen. The only reprieve came in the form of two weekly movies, shown on Thursday and Friday nights. This programming, however, left no room for personal interests or scheduling. As Omid, a film enthusiast and former video user, explained to me, it was a "big risk" to organize one's schedule around these timetables and then discover that the movie was either uninteresting or so heavily censored that it no longer made sense. In contrast, the videocassettes that a dealer brought to his family's home once week offered everything that television at that time did not: the right to pick what he wanted to watch and the freedom to enjoy it on his own time. He and his family would watch movies on videocassette at night, when the bombings tended to be more frequent. The movies were an opportunity to escape the war and its constant, oppressive threat.

For Omid, the most exciting part of his encounters with the process of distribution was not the video dealer's arrival or even putting a newly procured cassette in the video player, but rather the moment when the video dealer opened his metallic Samsonite briefcase (figure 3.2). It was in that moment that an entire week of escape unfolded in front of him. Likening the dealer's briefcase to a small treasure trove radiating golden light, he described how his family descended on the piles of videocassettes, quickly scanning for new releases they had heard about but also digging for new gems in the dealer's collection. Labels, Omid claimed, were everything. Because all of the videocassettes were stored in the same generic cases, the labels provided the only entry point into the content. All of the videocassettes were carefully placed with the label facing up to facilitate the selection process. According to Omid, each time the video dealer opened his briefcase, a minute of silence would ensue as the entire family read the titles and genres of each of the videocassettes. Then a flurry of opinions and questions would erupt about the five videocassettes they would select.

The labels video dealers produced symbolized the role they played as intermediaries between a new media technology and its consumers. The video dealer who visited Omid's apartment once a week translated all of the titles into Persian, sometimes taking liberties with the translations to help clarify movie plots, emphasize some noteworthy feature, or even misdirect the consumer. For example, Omid remembered that rather than translating the title of Oliver Stone's *Platoon* (1986) into the corresponding Persian term *jukheh*, he rendered the English title into the Persian alphabet. Intrigued by

Figure 3.2
The documentary *VHS Diaries* shows a briefcase full of videocassettes ready to be rented.
Source: VHS Diaries (dir. Niyaz Saghari, 2018). Courtesy of Niyaz Saghari.

the strange title—although uncertain of what it meant—and convinced that any Stone film is worth watching, the family rented the cassette for a week. In the process, however, they broke their one rule: no war films. Like many families, they actively tried to avoid movies about war, since they were living the realities of war every day. This was no easy task given that the 1980s produced a number of popular movies that centered on war. In the end, *Platoon*'s anti-war message sufficed to convince the family to continue watching a movie that broke their only rule. They couldn't help but wonder, though, if the video dealer had purposely mistranslated the title in order to get the videocassette in circulation when so many people were avoiding war movies.

As Omid's memories illustrate, videocassette labels guided consumers through the process of selection. They provided a snapshot of information, from which customers exercised their ability to choose what they wanted to consume. In fact, people began to associate the basic facts of a movie—as the video dealer represented them on a label—with the very possibility of selecting media. At the same time, the label was merely a material mark of the work that video dealers did mediating videocassettes for their consumers. This intermediary role made the work of the video dealer necessarily creative and cultural. Through labels, summaries, critiques, and suggestions, the video dealer played a part not only in how movies circulated in Iran in the 1980s and 1990s but also in how people made sense of them.

While for most dealers, labels were little more than a technicality—something thrown on at the last minute—others made them into an art form. Perhaps no dealer in Tehran was more famous for his labels than Mr. J, whose name came up in a number of my interviews, each time with mention of his impressive system of color-coded labels. A mysterious figure who lived far outside of the city where space was of less concern, Mr. J supposedly had one of the largest videocassette collections in Iran. He operated in the late 1990s and early 2000s, after the video ban had been lifted, and specialized in arthouse films, often obscure ones. Mr. J was a sound technician by training, and he viewed his videocassette dealing as an extension of his work with the film industry. In 2003, as part of a series of arrests of those who worked in or adjacent to the film industry, Mr. J was caught with his impressive stock of videocassettes. He was thrown in jail and tortured, an experience that profoundly affected him.

In her documentary *VHS Diaries*, London-based filmmaker Niyaz Saghari tracks down Mr. J, who had been her former video dealer. In a powerful sequence, she asks him to reenact his work as a video dealer (figure 3.3a). Unlike other reflections on the video dealer in Iran, Saghari doesn't just focus on how the video dealer transported videocassettes. She also dedicates attention to how he prepared the cassettes for distribution. We watch as Mr. J carefully prepares labels for his videocassettes. Stocked with a blank sheet of labels and a rainbow-selection of pens, he writes the genre title, director, and release date of the films in measured handwriting. As he writes, he narrates what he is doing in a gentle, thoughtful voice. These are the labels for which he had become famous (figure 3.3b).

Through labels and other designations, video dealers created a framework of knowledge for understanding films. In the absence of a formal industry for videocassettes, video dealers became responsible for framing the meaning of a film. Scholars of film genre have shown that the way in which marketers identify a film affects how viewers receive it.[41] When viewers sit down to watch a movie labelled as horror, for example, they evaluate it according to what they understand a horror film to be and what it should accomplish. Was the movie scary enough? Did it reveal the

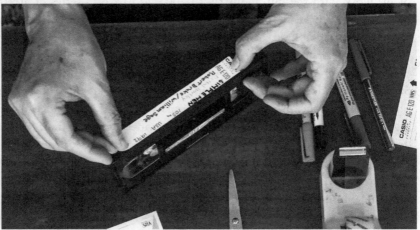

Figure 3.3a, b
In the documentary *VHS Diaries*, esteemed filmi Mr. J reenacts his former work
as a video dealer. He (a) takes to the street with his trench coat and Samsonite and
(b) creates detailed labels that helped frame his video collection for his customers.
Source: VHS Diaries (dir. Niyaz Saghari, 2018). Courtesy of Niyaz Saghari.

villain too soon? Did the soundtrack effectively signal danger? In this way, genre theorists argue, genres themselves do not remain fixed categories but rather are remade through modes of production, distribution, exhibition, and reception. This understanding of genre is productive here because it shows the role that distributors play in determining a film's meaning. Mr. J's careful labels were more than just an act of archiving information; they also framed a meaning for film. Even dealers who did not meticulously document information in their labels helped establish the meaning of films through the summaries and critiques that came out of their conversations with their customers.

Video dealers had to be creative in their work curating movies in part because it was physically impossible to store and transport many videocassettes at a time. The material constraints of the video underground also required dealers to be innovative in the very act of recording and duplicating tapes. Video dealers were confined in a number of ways, especially in terms of space, financial resources, and equipment. They could not simply record all of the movies in circulation in Iran at a given time. Especially compared to their digital counterparts, videocassettes were a bulky technology, even Betamax cassettes, which were called "small cassettes" (navār-e kuchak). There frankly wasn't space for all of these movies. More pressing, however, was the relative scarcity of videocassettes in the country. At various points, the government explicitly banned the importation of blank videotapes.[42] This created an economy in which video dealers constantly taped over content in order to keep their stock of cassettes in circulation. There weren't enough tapes to go around, which made them valuable commodities. As a result, video dealers had to be selective about which movies they kept.

Hussein, the telecommunications lineman who moonlighted as a video dealer, told me that he often had to make difficult choices about which movies to keep and which ones to record over. He explained that the market wasn't as competitive as one might imagine. When a new movie hit, video dealers would not go out of their way to keep it from one another. On the contrary, video dealers often knew each other and whenever someone in the group got a new movie on videocassette, he would pass it around and they would all take turns copying it. Unfortunately, however, they could not always copy everything they got access to, because they did not have enough videocassettes. For example, Hussein remembered the dilemma he faced when *Back to the Future* (dir. Robert Zemeckis, 1985) came to Iran in the late 1980s. It was the talk of the town, and Hussein could not afford not to have it in circulation on several cassettes so that multiple customers could rent it out at once. This required that he tape over some of his favorites, including *Ghostbusters* (dir. Ivan Reitman, 1984), which had arrived just a year before. Although Hussein hesitated, he decided in the end to serve his customers' interests over his own. Pressing the record button, he prayed that he would be able to find a version of *Ghostbusters* at another time in the future. How dealers made these selections had to do with the identity they cultivated for themselves or the kind of niche they tried to fill. Hussein catered to a client list interested in the biggest hits of the day.

Video dealers exploited the technological affordances of video. They reused cassettes, recording and rerecording over existing content for as long as the delicate magnetic tape allowed. They also worked off of versions of movies that were far removed from the studio-manufactured videocassette. As a result, the quality of a movie's sounds and images often deviated significantly from one another and from those of the studio-manufactured cassette. In fact, with one rare exception, none of the people with whom I spoke—dealers and consumers alike—had ever seen a studio-manufactured cassette. Once again, it would be very easy to talk of an original and a copy and to position the latter as inferior in every way. The difference in quality, however, doesn't have to evoke the rhetoric of deficiency, which emphasizes what these videocassettes lacked. English might favor a term like "the original" to describe the studio-manufactured videocassette that gave way to hundreds or even thousands of reproductions. I find the Persian *navār-e mādar* or "mother cassette" more productive here. Such a designation dismantles the hierarchy set up by the rhetoric of the original and the copy. In the mother cassette scheme, video dealers spurred future generations of cassettes, which did not necessarily have to index an original directly.

Through their recording practices, video dealers created cassettes distinguished by unique qualities—from warped sounds to frequent tracking to interruptions of white noise (*barfak*). These qualities—the marks left behind as a result of the work video dealers did recording content onto the tape's magnetic ribbon—became part of the experience of watching videocassettes in Iran. The marks, however, were hardly considered deficiencies. In fact, most people remember them fondly. At worst, my interlocutors expressed ambivalence about the unique qualities of these second-, third-, and even fourth-generation videocassettes, such as Nahid, who emphasized that "we watched the entire movie no matter how bad or how good the quality was, and some of them were really bad."

More often, however, my interlocutors found pleasure in these distinguishing marks. They were embodiments of the cycles of reproduction and the limits of videocassette technology, both of which shaped the video dealer's work. During interviews, I asked my interlocutors to recall a videocassette that they watched despite its poor legibility. In many instances—and to my surprise—they named the same movie that they had mentioned just minutes earlier when I asked about their favorite videocassette. It is significant that when asked to look back, Iranian video consumers conflate their favorite cassette with the one they found most difficult to watch. While it is impossible to know whether or not these videocassettes were actually people's favorites at the time or are just the most memorable years later, what we do know is that the people with whom I spoke did not associate illegibility with frustration, disparagement—or any negative emotion, for that matter. Instead, they spoke about the videocassettes' scratches and blurs with affection.

These small disruptions became part of the pleasure of watching movies at home. To return to an earlier example, when Media Farzin received a copy of *Last Year in Marienbad* from her video dealer, she watched it several times, each time struck by the fact that "somehow the stark, high-contrast cinematography was enhanced rather than diminished in the blurry transfer to videotape."[43] As this example illustrates, the seemingly uncomplicated consequences of the video dealer's recording practices became part of the experience of watching these movies—something to analyze or enjoy. Far from subordinate to some imaginary original, these cassettes helped render new meanings and new emotions for their viewers.

It is important to bear in mind that the videocassettes circulating in Iran in the 1980s and 1990s rarely housed just a single movie. Video dealers recorded additional content after the feature, making sure to fill up every available millimeter of magnetic tape. After watching a particular movie, viewers would likely encounter a teaser, music video, or show that the video dealer had carefully selected, like a small dessert to punctuate the main dish. Sometimes this additional content consisted of a brief clip from another movie that the video dealer had in stock. These teasers, usually just the opening scene, doubled as advertisements, a way for video dealers to get viewers hooked so they would continue their business the next week, or perhaps even sooner, depending on their enthusiasm for what they had seen. More often, however, the additional content consisted of some type of musical programming. This explains how many Iranians fell in love with Michael Jackson's "Thriller," or how those Iranians too young to remember life before the revolution discovered the country's premier diva, Googoosh, who—like all women—was banned from singing in public after the establishment of the Islamic Republic.

The filler material at the end of the cassettes also gave viewers their first taste of musical programming produced by the Iranian diaspora in Los Angeles. These programs, when featured at the end of a videocassette, were called *shows* in Persian and linked Iranians in Iran to the large diasporic community outside of the country. Housing the largest community of Iranians outside of Iran, Los Angeles—sometimes referred to as Tehrangeles—has been a production and distribution hub for Persian-language media since at least the mid-1970s.[44] Videocassettes played an important role in this industry, connecting different Iranian communities around the world before satellite dishes or online sharing. This industry catered to the diasporic community specifically, although its media content also found resonance with Iranians in Iran. Before satellite dishes arrived in Iran in the early 1990s, allowing Los Angeles production companies to beam their programs directly into Iranian homes,[45] content arrived to consumers in Iran on magnetic tape, where they occupied the last few minutes of a videocassette. For many of the Iranians with whom I spoke, these shows generated just as much excitement as the movies themselves. Some remembered exactly what show and movie were coupled together on videocassette. Others claimed that they would fast-forward to the show first, lest they die of anticipation. After all, they knew which movie they had rented, but had no idea what awaited them at the end of the cassette.

These shows meant so much to Iranian viewers because they offered an alternative to the mediascape in Iran. Leila, for example, explained that the shows at the end of videocassettes surprised her as a young child because they challenged the categories she thought she understood. Even from a young age, she knew that videocassettes were illegal objects. Through these illegal cassettes, she and her family watched recent Hollywood hits but also found delight in old Iranian movies from before the revolution. To see contemporary Persian-language music videos—even in the style of Michael Jackson, Devo, or Madonna—challenged what she understood as old and new, Iranian and non-Iranian. It startled her to imagine a place where Iranians could produce the provocative shows that ended up on some of the videocassettes her family rented.

By adding content like teasers and music videos to the end of these cassettes, dealers curated unique experiences for their viewers. I asked video dealers how this practice originated, and they all agreed that it initially had to do with the scarcity of the videocassettes themselves. Because of the limited number of cassettes in the country, they felt pressured to make the most of them. To allow a videocassette that was only two-thirds full onto the market seemed like a waste. Video dealers also told me that they understood that they were operating in the weak wartime and postwar economies. Despite the relatively cheap cost of video rental, videocassettes were still a luxury item for most people. Video dealers thus wanted to ensure that their customers got the most bang for their buck.

The dealers also enjoyed pairing movies with the end-of-the-tape content. Many video dealers looked for connections between the feature movie and the musical content at the end of the videocassette, adding Michael Jackson's zombie-inspired "Thriller" to the end of a George A. Romero *Night of the Living Dead* (1968) movie, for example. Hussein, however, told me that he preferred to surprise his customers with musical programming he added to the end of cassettes. For example, he told me that he punctuated Miloš Forman's biopic about Mozart, *Amadeus* (1984), with a video of Donna Summer performing "Hot Stuff." Hussein admitted he found it amusing to pair a classical maestro with the queen of disco. He laughed heartily when he imagined what his customers must have thought when the synthesized introduction of Summers's song began to play, especially following *Amadeus*'s somber ending. When I pushed him further, he also confessed there was a critique to be made about the state of music in the late 1970s and 1980s, although he refused to clarify his criticism. Nevertheless, Hussein's work adding content to the end of videocassettes was an act of interpretation. By carefully pairing a movie with additional programming, he and other video dealers offered a frame for understanding the movies, leaving their own interpretations inside the cassette.

The marks that video dealers left behind were not always just limited to piecing together existing content alone. They occasionally intervened more directly in the production of the movies they circulated. Specifically, video dealers sometimes took on the responsibility of dubbing the movies they entered into the video market. Because most videocassettes entered the market through informal means, they were not equipped for global distribution. On a practical level, this meant that they were not dubbed or subtitled in Persian. Maani Petgar, the filmmaker who briefly owned a video store in 1983 before relocating his business underground, told me that his English skills were essential to his work. As he explained, "During that time, my introductions to the films were important. Much like now, not that many people knew English well." This was in contrast to his work before the ban. The films permitted for release on video in Iran between 1980 and 1983 would have been dubbed and often times censored.

The movies that circulated on videocassette during the ban existed outside of the formal institutions that would have supported dubbing. While dubbing was integral to the video industry before the ban, no formal system for dubbing or subtitling movies on video existed after the ban. While many everyday users recalled watching movies in their original language with no subtitles, they also remembered that a surprising number had been dubbed into Persian. This work was likely accomplished by video dealers or other people closely associated with the underground video industry in the 1980s and 1990s.

While none of the video dealers I interviewed dubbed movies themselves, several consumers remembered watching movies that had been haphazardly dubbed by their dealer. For example, one of my interlocutors, Omid, recalled watching a version of Steven Spielberg's E.T. the Extra-Terrestrial (1982) that had been dubbed by his video dealer in the early 1990s. According to Omid, it was clearly not a professional job; the dealer had merely recorded his voice over the movie's soundtrack. Lacking the devices or skills necessary to mix the recordings, his dubbing acted more as a description or commentary for the images, rather than an actual translation of the movie's dialogue. Nevertheless, Omid—who had watched E.T without dubbing once before—appreciated the video dealer's commentary, happy to understand the strange alien movie at last.[46]

While not much evidence exists to connect individual dealers to dubbed versions of movies, this kind of creative work has remained important for Iranian audiences. In 2005, Film magazine conducted a large-scale survey to determine the best dubbed film of all time. Gone with the Wind (dir. Victor Fleming, 1939) came in fifth place,[47] a surprising victory for a film released almost seven decades prior and banned in Iran for nearly forty years. Even more surprising was that the dubbed version of the movie that most people remembered was not from subsequent theatrical releases of the film in the mid-twentieth century. Instead, the standard version had been produced and released on videocassette in the early 1980s.[48] According to Khosrow Khosrowshahi, who oversaw the videocassette dubbing of *Gone with the Wind*, a former classmate approached him about dubbing a copy of the movie that he had purchased for the video market. This classmate had made a comfortable sum of money in jewelry and became an early player in the underground videocassette distribution game. As Babak Tabarraee has shown, the many dubbed versions of *Gone with the Wind* in Iran represented distinct forms of authorship, each with its own critical discourse.[49] This view of dubbing as a form of authorship reinforces the idea that a movie's distribution imbues it with certain values. It also positions the underground video industry as creative and productive rather than simply a factory for copying and distributing movies.

Video dealers were curators and tastemakers. They did far more than secure copies of movies on video and deliver them to people's homes. They also had to prepare the videocassettes for distribution. In this way, they directly intervened in their content and exercised a degree of creativity. These were small, one- or two-man operations, so the same people who delivered the videocassettes were also responsible for managing their own supply. Through these management practices, they produced videocassettes that equaled unique viewing experiences unto themselves rather than mere copies of the original. Specifically, video dealers selected and recorded content; produced labels; added teasers and other material to the end of movies; and sometimes undertook the incredible work of dubbing foreign films. Each of these tasks required expertise knowledge, technical skill, and, yes, creativity, too.

A VIDEO DEALER FOR THE DIGITAL AGE

On one spring day in 2016, I took a break from interviewing former video dealers and videocassette users in order to eat at one of my favorite restaurants in Tehran. Offering an admittedly expensive take on the traditional dish *dizi* (meat stew), the restaurant located off of Iranshahr Street in central Tehran was crowded on that Friday afternoon. Throngs of people had gathered up and down the street, all waiting to be called for a table. My friend and I settled in for what would undoubtedly be a long wait. As we happened upon a shady spot, a man walked up to the large crowd that had gathered and laid down a blanket on the sidewalk. There he arranged scores of DVDs in careful rows. Each DVD was in a plastic sleeve with a paper copy of the movie's DVD jacket, film poster, or some other promotional material. With a captive audience, the young man began advertising his collection. Not unlike the traditional storytellers in Iran who retell epic tales like the *Shāhnāmeh*, this modern-day video dealer performed animated plot summaries of the movies he had available for sale—everything from *Avengers* and *Hunger Games* to *Fifty Shades of Grey* and *Pitch Perfect*. The crowd was transfixed, laughing at all of his jokes and shouting out requests for more information, especially about *Fifty Shades of Grey*, which he promised was completely uncensored. More than once, I saw the restaurant staff impatiently call the names of patrons who were too busy watching these movie performances to notice that their table was ready. The marketing strategy proved effective: numerous people bought DVDs from him over the course of thirty minutes or so.

I, too, found myself entranced by his gripping account of *Gravity*, when a hush fell over that street block in Iranshahr. A police car pulled up, and two officers started packing up the video dealer's DVDs. As they forced the man into the back of their car, the crowd began to protest. One person yelled, "Leave him alone! He's not bothering anyone." Another said, "He didn't do anything wrong. We were just having fun." The police were unfazed by the people's pleas. They carefully set the man up in the back seat of their car before driving off. The stunned crowd stayed quiet for a few moments, before someone chimed in, speculating that the police were probably just going to circle the video dealer around the block before letting him go. At worst, the man claimed, they would confiscate his stock of DVDs. The crowd's energy soon returned, and I heard small groups of people around me talking about movies while waiting for their tables. Sure enough, a few minutes later, the same video dealer walked down the sidewalk on the other side of the street. The crowd cheered, and, empty handed, he waved, before turning the corner out of sight.

An anecdote like this one pulls into focus the long history of video distribution in Iran as much as it reiterates the creativity implicit in the work of informally circulating media. Video dealers are still alive and well in Iran today, although video distribution as a form of labor has grown much more diffuse. From street vendors who specialize in DVDs to video store own-ers with a stock of illegal films behind the counter, the work of informally distributing media now intersects with a wide array of technologies, plat-forms, and industries. Digital technologies have made duplicating movies easier, while also maintaining better picture and sound quality. The arrival of the internet in the late 1990s has meanwhile handed Iranians access to almost everything imaginable. That said, internet plans in Iran are based not only on a fixed timeframe (e.g., one month) but also a fixed amount of data (e.g., 50 GB), so it does not always make much sense practically or finan-cially to stream content online. As a result, urban Iranians—in my obser-vations, at least—continue to purchase more DVDs than their American counterparts.

Because movies continue to matter as material objects—as discs and flash drives—an infrastructure remains in place for informally circulating them throughout cities. Indeed, as I argue throughout this book, the ban on vid-eocassettes in the early 1980s created a system of informal media distribution that continues to operate today. Even as the technological, industrial, and political conditions have changed, the basic modes of informal distribution have not. Video dealers continue to play an important role in how Iranians engage with and imagine movie culture. As such, they remain visible and beloved figures in the landscape of Iranian cities.

HOME VIDEO: PLEASURE, PERIL, AND PRIVATE SPACE

It was the spring of 1999, and Navid was sixteen years old. He had always felt a little bit like an outsider. Compared to his classmates, he was a more devoted student. Not only that, he was a fan of the Esteqlal soccer team, while the other students all preferred Persepolis, its rival team. Just as today, soccer mattered to Iranians at the turn of the twenty-first century. It even shaped school politics. Worst of all, Navid had never seen James Cameron's masterpiece *Titanic* (1997), which seemed to be all anyone could talk about. Around school, *Titanic* was infamous not only for its provocative sex scene but also its three-hour running time, which exceeded the length of just one videotape. The mammoth *Titanic* thus voyaged through two videotapes.

One day after school, Navid happened upon an opportunity to level the playing field. Nima, the biggest bully in school, had left his backpack wide open as he chatted with some friends nearby. Two *Titanic* videocassettes sat comfortably on top of his stack of books. They were practically begging to be stolen. Without even thinking, Navid snatched the cassettes and ran home, never looking back. He immediately watched the three-hour movie and, already a budding film buff, found the special effects impressive but felt the plot left something to be desired. Steamy car scene aside, he didn't quite understand all of the fuss.

As soon as the film was over, a wave of guilt and fear washed over Navid—not because of the provocative scenes he had just watched, but rather because he remembered how he had gotten the videocassettes in the first place. Certainly, Nima had already realized that his copy of *Titanic* was missing, and you can imagine how much cultural capital a film like that carried at an all-boys school in Tehran in the late 1990s. Panicked, Navid called his friend Shahin, who agreed to find a way to return the videocassettes to Nima's backpack—after he had re-watched a few scenes himself, of course. They agreed to meet at a nearby park so that Navid could hand off the cassettes. It was already late, and Navid snuck out in his pajamas, the *Titanic* videocassettes in a clear plastic bag.

Navid found Shahin pacing in front of a bench. He walked up to his friend and extended the bag forward. Just as Shahin took hold of it, sirens filled the Tehran night, assaulting the boys' eyes and ears. They had been caught. When the police saw the *Titanic* videocassettes, they dragged the boys to a station miles away. As the officers threatened jail time, Navid cried and pleaded shamelessly, worried that any kind of record would ruin his chances of going to a good university. In the end, the officers agreed to let the boys go under one condition: they had to destroy the videocassettes right then and there.

The officer threw the two cassettes on the ground and made the boys stomp them to pieces. Navid didn't mind. It wasn't like he was being asked to destroy *Annie Hall* or some other great film. When they were done, the officers pushed Navid out into the cold and told him to return home immediately. After a few minutes of standing there in his pajamas, defeated, he began the long trek back home. A small piece of plastic stuck in the treads of his shoes crunched with every step.

Navid's story is emblematic of the ways in which the values attached to videocassettes transformed as they moved through different spaces—from the schoolyard to the home to the police station. Videocassettes might have been precious capital in teenage social exchange at school and a source of secret pleasure in the home. But in a police station's interrogation room, videocassettes were the embodiment of crime and decadence, even worthy of violent destruction. Navid's experience with video was not unique. Throughout the 1980s and 1990s, Iranians juggled different subjectivities as they negotiated a multitude of public and private identities. In much the same way, videocassettes absorbed the myriad social and political forces that determined the spaces they occupied. In this chapter, I argue that the underground video network—which operated at the nexus of public and private space—helped Iranians negotiate the different identities they had to perform. As I will show, videocassettes charted Iranians' broader relationship to the state and the encroachment of the government into their private lives.

Oral history interviews have been central to my understanding of the ways in which the video infrastructure cut across people's private spaces and into their lives. As media historians have shown, it is particularly difficult to study the lived experiences of domestic technologies.[1] Not only are such experiences vast and diverse, but available sources rarely attest to the private practices that people develop within their homes. Oral history offers an exciting solution. Of course, oral history interviews do not yield objective facts, but they do teach us about narrative strategies and the patterns that structure people's memories.

My corpus of oral history interviews reveals that people tend to tell two kinds of stories about home video technology. I call these narratives the perils and the pleasures of video. In such accounts, we can see that people's memories often organize themselves around two poles: either instances of getting caught or moments of intense enjoyment. By studying these extremes, we stand to learn a great deal about what it meant for everyday Iranians to welcome video technology into their homes. Taken together, stories of peril and pleasure unravel how the experience of videocassettes in Iran in the 1980s and 1990s was part and parcel of other transformations at the time. People adopted videocassette technology in reaction to these changes, especially the blurring of the boundaries between public and private space.

By examining the triangulation of home video, state power, and private space, this chapter joins a body of scholarship that seeks to theorize the experience of watching movies at home.[2] Barbara Klinger has suggested that the home is "host to an array of film cultures, each characterized by an elaborate set of aesthetics, viewing modalities, and pleasures."[3] Similarly, Caetlin Benson-Allott has called for more critical accounts of the moment when "going to the movies started to mean staying at home."[4] Studying the "home as a media hub" requires that we examine how media technologies operate at the threshold between public and private spaces.[5] In Iran, underground video provided a foil to public spaces, which were saturated with the power and dangers of the state—including war, surveillance, and regulation. This relationship explains why videos were a source of pleasure but also of peril. The pleasures of video were won through subversive means punishable by the state, a threat that was always present.

A PRIVATE RETREAT

As war and revolution raged outdoors, the home became an attractive space both for refuge and for the consumption of media. The violence of war, political instability, and the surveillance of public spaces in Iran meant that, throughout the 1980s and 1990s, people retreated into their homes, where they could lead private lives separate from the state. With few attractive public spaces for leisure and entertainment, the home became an integral part of people's social lives. Video technology was central to the practices and leisure-time rituals that people formed within the domestic sphere. Iranian society's retreat into the home in the late 1970s and early 1980s coincided with analog video technology's arrival. One man who was in his twenties in the 1980s recalled, "You could say that was the golden period for video, because we would have parties to sit together at home and watch movies. You'd rent an entire suitcase of movies on video for twenty-four hours. You had twenty-four hours, but there were way more movies than that at your disposal. You usually had at least seventy-two hours' worth of films. . . . This was the difference with cinema: the power to choose. So, you'd host a party and sit together and watch as many movies as possible."[6] A memory like this brings into focus the two ways in which video culture facilitated the integration of social life into the domestic sphere: (1) by offering users the ability to choose the parameters of media consumption, including what they watched, when they watched it, and with whom; and (2) by fostering occasions for different kinds of people to come together in close proximity within the home.

The home constituted a critical node in the vast underground network. The video infrastructure facilitated the transportation of cassettes through public spaces to allow for their safe arrival to homes. In this rental system, however, a videocassette's journey did not end in these domestic spaces. Rather, it occupied the home for several days before circulating through public and private spaces once more. The home was a stopover more than the final destination—an intersectional point in a complex web of movement, material, processes, and labor. The stories that take shape around such a point constitute what Nicole Starosielski calls "nodal narratives," which expose "the conflicts, contestations, and negotiations that shape systems on the ground."[7] By acknowledging the home as a node, we can appreciate how videos became indistinguishable from the pleasures of the domestic sphere, allowing people to stay home and stay entertained.

Videocassettes were part of a larger retreat of public life into domestic spaces. A survey of university students from 1993 sheds light on viewing habits and motivations for adopting video despite its illegality. In these accounts, the failure of the public sphere necessitated that the home take on the role of the university. One communication sciences student stated,

"It would be ideal if our university had a space for watching videos, and students could . . . discuss what we had seen after watching it together. But such an opportunity doesn't exist, so we have no other choice than to watch movies at home alone or with a group of friends for fun." Another student, who studied filmmaking, made a similar point: "Of course, because it is related to my work to watch movies, and a lot of them, I have to use video. Every year there are useful movies that don't make it to our cinemas. This problem and deficiency force me to seek out video."[8] Statements like these show how videocassettes facilitated a retreat indoors, where public life continued within the walls of video-equipped homes and apartments.

As the home became the site of an increasing number of social and leisure activities, videocassettes integrated into a new form of domestic life where public and private interactions often took place concurrently or at least in the same space. For many Iranians, videocassettes may have represented the ability to stay in but they didn't necessarily render movie-watching into a solitary pastime. My interviews revealed that watching movies on video was a communal activity, usually involving family. Especially during the ban, in the 1980s and early 1990s, extended families often only had one video dealer, weekly subscription, and video player. The entire family would gather to watch the rental videos together. As one woman remembered, "Family members would come to our house and sit like it was a movie theater. At that time, our friends and family didn't necessarily have video players at home, so they would come to our house and we would all sit together."[9] In such instances, the home became the movie theater, as family members and neighbors lined up chairs and sofas in rows in front of the small television screen. A recent essay on videocassettes in the popular magazine *Andisheh-ye Puyā* put it particularly well: "the invigorating, euphoric experience of watching a film, which had been stripped from the movie theater, was now being produced again in the home through video."[10]

While watching a movie at home may have replicated some of the same sensations as the movie theater, it was also a unique experience unto itself. One of the key differences was the degree of control that video technology provided its users. For example, Farhad—who was a young father of two in the 1980s when he first encountered home video technology—told me that videocassettes represented "not needing to go to the movie theater. I could sit at home and watch exactly what I wanted . . . it was completely in my control." For many Iranians like Farhad, videocassettes reified the desire to stay home, where control was not only possible but also prioritized. He continued, "Outside they could dictate what I wore, how I styled my hair, and where I went, but at home with these [videocassettes] I was in charge of my own fate." The control that video afforded Iranians overlapped with their ability to perform private identities at home, a privilege that public space under the new Islamic Republic did not grant them.

Indeed, it was the video player's remote control that gave body to the control that many Iranians aspired to within their homes. As Benson-Allott writes, "We have reserved the name *remote control* for those gadgets that express our dreams and anxieties about how we relate to radio, television, and other audio-visual media."[11] In one of my interviews, the remote control became the device through which my interlocutor, Houshang, remembered his early encounters with video. He recalled, for example, the ability to pause when extended family stopped by unexpectedly; to rewind when his daughter cried drowning out an important line; or to fast-forward through the parts he thought were inappropriate for his two young children. Houshang found control over his leisure time in the liberties that home video technology provided. Not only did videocassettes offer the option of staying home but also the technology itself yielded control that public life did not: he could pause, rewind, and fast-forward according to his own needs and desires.

The experience of watching movies at home with extended family was possible in part because the mid-twentieth century had seen a restructuring of urban architecture in many cities across Iran. The traditional Iranian home consisted of a large courtyard surrounded by individual apartments that each housed extended family, so that grandparents, aunts, uncles, and cousins all lived in the same home. In cities like Tehran, a growing urban population put pressure on limited space throughout the twentieth century. Large apartment buildings replaced traditional courtyard homes. The principle of communal living remained for many middle- and upper-class urban dwellers, but space was now divided vertically instead of horizontally. Many apartment buildings were owned by families, with different members each occupying a single floor. Even if families lacked the luxury of owning and occupying a single building, they often lived nearby, sometimes in the same neighborhood. It was easy for families and even friends to gather in the evening to watch a movie on videocassette all together.

This kind of communal viewing was not entirely without precedent. During my interviews, I heard several instances from the 1970s and early 1980s, before the advent of videocassettes, in which families would gather to watch movies on film. Usually one person in the family owned a film projector or *āpārāt* in Persian, which allowed the entire family to watch rented or purchased movies on 16mm film. According to my interlocutors, these 16mm films were often the source for early movies on videotape. Watching films at home was not a large-scale phenomenon—not nearly as widespread as home video technology would become—but it was common enough that many of the people with whom I spoke had experienced or heard of it.[12]

Yet for most families the possibility of watching movies together began with the adoption of videocassette technology. Video technology played a particularly important role during the war, which kept families at home with nothing to do. Figures show that every month throughout the 1980s approximately 11,500 families in Tehran would gather to watch movies together.[13] One man remembered, "When the war started, people would go home early. Suddenly, hanging out together [*dowr-e ham jam' shodan*] at home became a regular occurrence. Women of the house didn't mind; they were happy to have their husbands at home at night. But spending time together required some sort of entertainment. Dealers hit a stroke of genius when they began circulating films on video to suit the tastes of any person."[14] Such a description shows how the realities of war helped create new social practices at home. Media often represent war in spectacular terms—as constant bombardment, violence, terror, and excitement. But the everyday experience of war is very different. For those living it, war is often marked by idleness and boredom. As one of my interlocutors, who was a child during the war, told me, "When people ask me about the war, I remember being bored. We weren't allowed to do anything. We were stuck at home most of the time. If it hadn't been for video, I would have died of listlessness [*khastegi*]."

As domestic life changed, newly adopted cassettes, movies, and players mediated gendered and familial relationships as they took shape within the video-equipped home. Sometimes complications arose as family members gathered together in the home with nothing to do. The fact that the man quoted previously needed to mention that "women . . . didn't mind" suggests the ways in which the gendered nature of the home was transforming.

The premodern Iranian home had specifically gendered spaces: *andaruni* (the interior) for women and *biruni* (the exterior) for men.[15] By the mid-twentieth century, however, the modern urban apartment didn't have such compartments. The entire interior became the domain of women, while men's leisure took place outdoors in coffeeshops, parks, cafes, and cabarets. In the 1980s, when war drove men home before nightfall and forced the entire family to spend significant amounts of time together, videocassettes helped establish the home as a social space for everyone.

As men and women gathered together for leisure, video gave them something to do and helped mitigate the pressures of having the entire family home at once. One of my interlocutors, Bahar, remembered the tension that came with being home together during the war. Although only a child, she witnessed her parents fight often. The instability of that period meant that food, gasoline, and other resources were scarce. According to Bahar, her parents struggled to keep the family together. She dreaded the days her father couldn't go to work because of bombings outside. More time at home together meant more fighting. It wasn't just that they fought about matters related to the war. Her father's time at home meant an invasion of her mother's space. Reprieve came in the form of a plastic videocassette. As Bahar recalled, "When we would get a new shipment of movies on video, I would calculate in my head how many hours of peace they would amount to. When we watched videos at home, life was good. The family sat together, my mom next to my dad. They would laugh and joke. It was a whole other version of family life that only existed when the video player was on." Bahar's memories position the home as a retreat from the terrors of public spaces. At the same time, they show how video mediated family relations as the domestic sphere became a place for social life, as well.

Positioning the home as an important site of media consumption expands the current scholarship on movie culture in Iran. Scholars have established the movie theater as an important site of media consumption. They have offered detailed descriptions of early theaters at the beginning of the twentieth century; undertaken textual analysis based on apparatus theories that make assumptions about the movie theater space; and performed ethnographic studies of reception with movie-theater goers.[16] The field has yet to acknowledge the Iranian home as a unique site of viewership, as a place where people watched and made sense of movies.

The irony of this lacuna is twofold. First, the scholarship on Iranian cinema outside of Iran has long depended on the distribution of Iranian films on various home viewing platforms, including videocassettes and DVDs. Second, home viewings were central to the very origins of cinema in Iran. The first public screenings were held in the backyard of a private home, while early Iranian film production was based in the court, where the shah and his family filmed themselves and then held private screenings in palaces.[17] Given this precedent, we as scholars should be asking what the experience of watching movies at home meant to Iranians, especially once the proliferation of home video technology made it a widespread possibility for many urban dwellers.

VIDEOTAPE INTIMACIES

Many Iranians who grew up on video look back and marvel at its small scale. It is somewhat of a historical truism that each generation imagines itself as the technological exception. For members of every generation, the dominant technologies of their youth made their experiences unique from both the generations before and after. This exceptionalism was certainly the case in Iran, where the generation that grew up with videocassettes understood its relationship to the space and scale of media as somehow different from previous and subsequent generations. As film critic Shamim Mostaghimi explains, only the videocassette generation came to understand media on such a small scale.[18] In a recent essay, he claims that the generation that came of age with movie theaters experienced cinema through high ceilings, big screens, and spectacular buildings, while the internet generation has subsequently enjoyed satellite dishes, larger and more powerful television sets, and a seemingly endless supply of media content. For Mostaghimi, the video generation was distinguished by the small scale of its devices: tiny TV screens in crammed living rooms and a limited supply of cassettes and players.[19] Media were scaled down in the 1980s and 1990s, and this afforded certain affects. Mostaghimi writes that the movie theater's big screen was like "screaming," but video was like "speaking softly in someone's ear. It was a whisper."[20] A description like this hints at the ways in which videocassettes fostered intimacy, and it invites us to imagine what those intimacies might have looked and felt like.

Home video technology was just a whisper because of proximity, because it demanded closeness. Small screens, poor sound quality, and distorted images meant that people had to be near their devices and one another. Videocassette technology foregrounded those material relationships. As a result, one kind of intimacy that developed during this time was between users and their media devices. Home video technology required that people interact with media as material objects in unprecedented ways—from carrying videocassettes to adjusting the television set to blowing into the video players, cleaning out dust. This was especially the case in Iran, where the underground infrastructure for rental meant worn cassettes and old, poorly maintained equipment. The relationships that developed out of this infrastructure were often as much bodily and tactile as they were emotional. In my interviews, people often described their affection for videocassettes by talking about their physical relationships with them. The vocabulary they used replicated the sensual textures of love and intimacy, as does Mostaghimi's choice of words when he likens video to whispering in someone's ear.

Sina, who was a teenager in the 1980s, remembered the time that he "wrestled" his family's old video player after it ate a cassette he had rented. Their dealer had already threatened to fire them as customers after the same VCR destroyed a tape several weeks earlier. This time Sina had to save the cassette or risk losing a good, dependable video dealer. As soon as he heard the irregular clicks and crunches, he sprung to action. He dramatically reenacted the scene for me during our interview. He mimed the push and pull of the videocassette as he pleaded with and cursed at the video player. He called it "djinn" and "ghoul," as though the incident were a scene in a great Persian epic like the *Shahnameh* and he was Rostam slaying the white demon. Miraculously, the video player loosened its claws and Sina was able to slide the cassette out. He even managed to wind the scattered magnetic tape back into the plastic shell. Relieved and not knowing what else to do, he hugged the videocassette and kissed it like it was his "lover" (*ma'shuq*). What's important here is Sina's casting of this story as an epic tale of love, one in which he foregrounds the materiality of media. Within a mere matter of seconds, the material of home video technology became a source of both frustration and affection for Sina. He had an intimate relationship with the cassette that far exceeds what one might expect. While Sina's recounting of this memory was exceptional, I heard many other stories that gestured toward this kind of intimate relationship with videocassettes.

In fact, some people loved their videocassettes so much, they would bring them to bed. Behnaz was only eight years old in the mid-1990s when she saved her small allowance for months to buy her first and only movie on video. By then, the ban on videocassettes had already been lifted. Buying movies on video was more of a possibility, although still not a widespread phenomenon. Disney's *Beauty and the Beast* (1991) had always been Behnaz's favorite, and she couldn't think of a better way to spend her money. With the help of her father, she bought the cassette from her family's video dealer. It wasn't a studio-manufactured cassette, but in a stern exchange, the dealer assured Behnaz that the movie was of high quality (*bā-kayfiat*). This was a serious purchase, and she wasn't going to leave anything to chance.

Once the cassette was in her possession, Behnaz refused to let it leave her side. On multiple occasions, she even defied her mother's warnings and brought the tape to school, where she would quickly show it to her friends while the teacher was turned to the board. What she remembered most about those first few weeks with *Beauty and the Beast* was bringing the cassette with her to bed every night and falling asleep with it in her arms. As Behnaz recalled, the irony of all the physical affection she gave her videocassette was that she only watched the movie a handful of times. In her estimation, the value of the videocassette far exceeded the enjoyment of watching its contents. She had seen too many cassettes meet their death in video players or through other careless accidents. On the few occasions when she had watched the movie, she demanded silence from her friends and family as she carefully removed it from the cardboard sleeve and gently placed it in the video player's terrifying mouth.

How might we explain Behnaz's special relationship to her *Beauty and the Beast* videotape? At first glance it might appear to conform to Baudrillard's assessment that people's attachment to objects grows out of a desire to materialize social status.[21] Videocassettes were rare in Iran and owning them even rarer. Behnaz took pride in having a videocassette of her own, even showing it off to her friends at school. We might also see resonance of Christian Metz's concept of cinema as a "fetish that can be loved."[22] It was Behnaz's love of *Beauty and the Beast*, after all, that motivated her to save her money and purchase the videocassette in the first place. Baudrillard's explanation may attend the social value of material objects, while Metz's captures the affective qualities of our relationship to media. The videocassette is both an object, a thing, but also a means of access to the sensations of media consumption. Braiding these two strands together and acknowledging the unique contours of media objects requires a supplementary vocabulary, one of video intimacy—of the affects produced through touching, holding, and embracing media.

Such intimacies are important because, at least in Iran in the 1980s and 1990s, the physical relationships that people developed *with* videocassettes were inseparable from the personal relationships they developed *through* videocassettes. Walter Benjamin gestures toward such an affect in his oft-quoted essay "Unpacking My Library," in which he claims that intimacy can define our relationship with media objects.[23] For Benjamin, this intimacy is achieved through collecting and ownership.[24] Indeed, collecting is a central motif in much of the scholarship on people's material relationships with media, especially video.[25]

Yet stories like Sina's and Behnaz's require that we move beyond collecting and ownership in understanding the intimacies that developed around media objects, since very few people in Iran owned or collected videocassettes. Hilderbrand offers a productive framework when he argues that video "tapes created social networks before 'social networking.'"[26] His analysis moves beyond the relationships developed between individuals and the media objects they possess. Rather than viewing a collection as a static library, he theorizes the gift economy created as video collections were shared and circulated. This economy, according to Hilderbrand, generates a range of intimacies among friends and even strangers.[27] In other words, video fostered a culture of closeness, as bodies and devices came together in the home.

These affective spaces were charged with a range of sensualities, touches, and emotions. One of my interviewees, Tahmineh, who was a teenager in Tehran in the early 1990s, shared with me the story of her first boyfriend—a relationship mediated by the closeness that home video technology fostered. Because her family was fortunate enough to own a video player, neighbors in the apartment complex often stopped by to watch movies. In particular, her family hosted weekly viewing parties with the family who lived on the floor below. Tahmineh had never really given much thought to the family's son, Ramin, who was about the same age. She had known him her whole life, but they went to separate schools and hung out in different circles. Tahmineh's thoughts about Ramin changed, however, during one of their weekly video gatherings.

As the two youngest people in the room, the privilege of picking the movie often fell to Tahmineh and Ramin. That day the two rummaged through a stack of four or five cassettes on the floor in front of the TV set. Tahmineh felt a shock of excitement when Ramin's fingers grazed her arm as he reached for one of the videocassettes. Although they usually disagreed about which movie to watch, Tahmineh gave into Ramin's choice without a fight. After hitting play, he squeezed on the couch next to her. It appeared he too had felt that charge of connection. When I asked Tahmineh what movie they watched that day, she said, "Oh God, I have no idea. I was so focused on his shoulder touching mine that I couldn't pay attention to what was on the screen." So began the torrid but innocent love affair, which played out almost entirely in the confines of those video viewing parties through the touching of shoulders, sideward glances, and those moments when they both reached for the same tape; their parents completely unaware of what was happening.

The circulation of videocassettes became a metaphor for love and relationships, or—as we saw in the case of Tahmineh—at least inspired new ways of dating. The increased surveillance of public spaces in the Islamic Republic foreclosed the possibility of public dating. At various points over the last forty years, unmarried men and women seen together in public have been harassed and interrogated by agents of different morality police. This drove the domain of love and dating into unexpected corners. We might consider, for example, the main characters in Shahriar Madanipour's novel *Censoring an Iranian Love Story*, who spend their first date in an emergency room lobby, the only place where they can sit together and have a conversation.[28] With these circumstances in mind, media played a role in how people communicated and experienced love.[29] Videocassettes were no exception, and my interviewees shared a number of stories of love and intimacy that took place in front of a TV while a movie on video played in the background.

For many, the experience of dating was intertwined with home video technology. Tara Bahrampour's memoir *To See and See Again* captures this entanglement. At one point in the memoir, the story of a suitor at a party in North Tehran in the mid-1990s becomes indistinguishable from a larger commentary on media access. In one passage, Bahrampour explains that "people in North Tehran watch a lot of American movies . . . via secret door-to-door vendors. One man carries them in the pocket of his coat; another dresses as a baker and places the video inside a hollowed-out cake, carrying it in a pink box to the door."[30] A few lines later, Bahrampour's suitor, Cyrus, asks if she has thought about his proposal of marriage. "In

the meantime," he suggests, "we can get to know each other through letters." Here the connection between the secret circulation of videocassettes
and the secret exchange of love letters is too obvious to miss.

Their conversation is interrupted when someone yells, "Hey! You've got
Queen!" before "slipping a video into the VCR. A long-haired guitar player
appears on the small screen, hammering out a frenzied solo." After the interruption, Tara and Cyrus continue their conversation. "'We'll have a letter
relationship, then?' Cyrus whispers."[31] Videocassettes and whispers dance
together in the air. The scene ends as Tara agrees to a letter relationship, and
then the newly minted couple goes to the balcony to survey the city below
while a helicopter searching for satellite dishes hovers above. *To See and See
Again* has very little to do with media but has a lot to say about growing up
in Iran at the time. It is no coincidence that Bahrampour frames this story
of dating, which only extends a few pages, with media infrastructures; the
circulation of videocassettes sets the love story in motion and satellite dishes
close the scene. Media technologies, especially videocassettes, nurtured the
home as a space where such intimacies became possible.

It wasn't just that videocassettes became implicated in the experience of
intimacy; they also offered many young Iranians knowledge of love and sex
to which they otherwise wouldn't have had access. Like other forms of public culture, sex and intimacy also retreated into the home in the 1980s. Sex
may have been a regular feature of Iranian movie theaters in the 1970s, both
in Iranian productions and the foreign imports that screened in Iran,[32] but
after the establishment of the Islamic Republic sex and love were expelled
from the movies. As anthropologist Pardis Mahdavi has shown, the state's
moralizing policies left young Iranians with very little knowledge about
sex and very little access to information about it.[33] Within this context,
videocassettes provided a crash course in what sex and love looked and even
felt like. As Linda Williams has argued, cinema is "the most powerful sex
education that most of us will ever receive."[34] In Iran people forged access
to this curriculum through videocassettes. For the generation that grew up
in the 1980s and 1990s, videocassettes were marked by their sexuality. This
was a theme in almost all my interviews with users who were children and
teenagers at the time. People's understanding of videocassettes as tied to
sexuality did not suggest a perverse relationship but rather one of curiosity and knowledge. However problematic and incomplete, videocassettes
offered a gateway to information about sex that the state denied them.

In this videocassette school of sex education, James Cameron's *Titanic* (1997) stood out. For many Iranians born in the 1980s, *Titanic* defined their experience with videocassettes and, perhaps unexpectedly, contributed to their understanding of sex. The significance of this tragic love story on video was not just evident in my interviews, where it almost always came up, but also in reported numbers from the time. According to estimates in Iranian newspapers, the underground circulation of *Titanic* on VHS in Iran generated nearly US $3 million in 1997 alone. This was more revenue than the entire national film industry earned that year.[35] Clearly widespread and popular, *Titanic* also embodied videocassette technology for my interlocutors for other reasons. For many of the people with whom I spoke, *Titanic* was the last movie videocassette they remember purchasing or renting. By the late 1990s, VCDs had begun to overtake videocassettes as the dominant home video technology. At this moment, *Titanic* represented the shortcomings of videocassette technology. The three-hour movie was so long that it required two cassettes, and the bulkiness of a videocassette—especially compared to a thin disc—became especially burdensome and apparent.

But *Titanic* also stood out in Iran—as it did elsewhere in the world—for its representation of sex.[36] Without hesitation, most of my interviewees who were adolescents in the late 1990s identified *Titanic* as the first pornography film they had ever watched. The first time I heard this classification of *Titanic* I was admittedly confused. How could my interlocutors discuss a movie like *Titanic* under the same umbrella as the hardcore pornography they would later admit to watching on video? As more and more people categorized the popular Hollywood movie as pornography, I came to understand that the term *porno* in Iran wasn't burdened by legal definitions or moral claims like in the United States. Rather, it was a stand-in for all types of representations of sex. What constituted obscenity seemed to matter very little.[37] More important in *Titanic*'s success was the fact that for many Iranians videocassette technology was wrapped up in their memories, experience, and knowledge of sex. Negar, for example, explained to me that before she watched *Titanic* at fourteen, she had only a vague understanding of what sex was but nevertheless associated it with negative qualities: pain, shame, fear. The sex scene in *Titanic* led to an epiphany for Negar, as she realized that sex could also be beautiful, pleasurable, and loving.

Ironically, the consumption of sexual images that *Titanic* represented for many Iranians was the state's biggest fear. As I've shown elsewhere, the laws governing video production and distribution after the ban reveal that the production and circulation of sexual images were at the heart of the state's antagonistic attitude toward video.[38] In 1995, the government issued master policies for video now that the ban had been lifted. These policies included a lengthy section about the punishment for producing, distributing, or possessing pornography (*mostahjan*), which is defined rather ambiguously as "any audiovisual work whose subject is the direct and explicit representation of sex." According to this document, the potential "degradation of generations" through the rapid duplication and distribution of pornography on video was "the reason for the ban [on video] in the years following the Islamic Revolution."[39] As a result, the punishments for distributing or producing pornography on video were quite severe. Even so, sexual images of all kinds made their way into Iranian homes via video.

In fact, it wasn't just Hollywood movies like *Titanic* that circulated images of sex in the country. The rise of videocassette culture also meant that explicit representations of sex—what in the United States might be called pornography—also moved through Iran on a relatively large scale. Just as elsewhere, the spread of pornography grew in tandem with videocassette technology.[40] Some video dealers in Iran specialized in distributing pornography; others carried a separate list of pornography to which customers could subscribe. The price of buying or renting pornography on video corresponded to the country of origin, with East Asian productions from Korea and Japan on the lower-end of the scale, followed by European and American pornography. The highest prices, however, were reserved for local Iranian productions.

Most of the people I spoke to admitted to watching pornography on videocassette at some point in the 1980s and 1990s. What stood out in many of these accounts was the fact that watching pornography—like watching other movies on video—was a group activity. These videocassette screenings included the full gambit, from groups of teenage girls secretly watching a pornography video while their parents were away to ritualized viewings with all the men from the family present, including uncles and male cousins. These screenings had little to do with sexual gratification. They were driven instead by curiosity and discovery, and the experiences and information shared between family and friends. I bring forward these stories of sex and pornography not to engage in larger debates about obscenity and spectacle or to question the kind of sex education that many Iranians received on video. Rather, I demonstrate how videocassettes fostered discussions and experiences of sexual intimacy in Iran in the 1980s and 1990s.

Somewhat unexpectedly, I heard more stories about watching pornography from women than I did from men. While most pornography is produced and distributed according to a logic that prioritizes male desire and pleasure, feminist scholars for nearly three decades have shown that female enjoyment is not necessarily absent in the consumption of pornography—even those examples specifically targeted at men.[41] Access is one of the most important conditions of women's enjoyment of pornography. In Iran, the foreclosure of public sexual culture for men after the revolution relocated it to the home, where women could enjoy it, too. The conservative policies of the 1980s and 1990s removed sexual culture from public spaces, where men had previously accessed them. These forms of sexual culture included cabarets, adult theaters, and even the mainstream cinemas deemed unseemly for women because of their sexual content. Once in the home, women had access to it, too. As Jane Juffer writes, "Pornography is domesticated when it becomes integrated into the routines of everyday life—not exactly rendered invisible, but managed by women who have achieved a degree of agency through a structured movement between sites of access and sites of consumption."[42] In the case of Iran, women's consumption of pornography on videocassette signaled the domestication of a public sexual culture.

Women's consumption of pornography on video and the tensions that exist around it are portrayed in Azadeh Moaveni's memoir *Lipstick Jihad*. Moaveni, who was born and grew up in the United States, moved to Iran in her twenties for two years in the late 1990s to work as a journalist. *Lipstick Jihad* narrates her experiences as an Iranian-American living in Tehran. One of her rituals in Tehran was watching movies on video with a "distant relative," a twenty-one-year-old woman named Mira. As Moaveni explains, "Western movies were banned, but everyone had a filmi, a video guy, who schlepped a trunkload of new films around to his clients' homes as a sort of mobile video store."[43] On one particular movie night, Mira "hit the play button, and a fuzzy image of two very white, very naked people appeared on the screen."[44] As it turned out, she had brought German pornography from the 1970s that day. According to Moaveni, Mira "expertly fast-forwarded to the next scene" and asked, "Why does he keep doing that?"[45]

She had selected that tape to initiate a discussion of sex with her American relative. As it turned out, her boyfriend had given her the cassette as a part of a "sex kit" so that he didn't have to "waste" their limited time together with "erotic tutorials."[46] The videotape's use as both a conversation icebreaker for two cousins and an educational resource between two lovers shows how home video technology fostered forms of intimacy in Iran at the time. Moaveni disapproves of this "accelerated, demeaning, furtive initiation into sexuality"[47] and calls Mira's boyfriend "seedy."[48] She concedes, however, that one of the greatest "causalities" of the Islamic Republic was the "casual coexistence between the sexes." In a gender-segregated society, Moaveni explains, "Being together involved sneaking away, into the dark corners of public parks, into the woods of the Alborz Mountains."[49] Videocassettes and sex were linked via their shared inhabitance of the underground, as private, intimate moments unfolded in the margins of public space.

Like Bahrampour's memoir, Moaveni's has almost nothing to do with videocassette technology. Yet it is not a coincidence that in both cases the brief descriptions of the underground video infrastructure appear alongside stories about love, dating, and sex in the Islamic Republic. These stories of love, sex, and intimacy were indistinguishable from the larger processes that provided access to movies on video to millions of urban Iranians. As we've seen, the underground infrastructure made possible a myriad of different videotape intimacies. Oftentimes the intimacies that people developed vis-à-vis video were with the technology itself, as they delighted in touching, treasuring, and trading their cassettes. At other times, videotapes mediated intimacy between people. In such moments, the underground simultaneously sheltered videocassettes and love, both of which had been rejected by the government's totalizing ideology. Far from being dark and criminal, the underground infrastructure pulsated with pleasure, passion, and proximity.

LESSONS AND LIES

As the underground video infrastructure snaked through public and private spaces, it instructed Iranians on the different subjectivities they could occupy in each realm. Never was the videocassette infrastructure only about private space or the intimacies that could develop there. Instead, it was a system of processes, labor, and bodies that transported media objects through public and between private spaces. Videocassettes meant different things and were evaluated differently in each of these spaces. As I just showed, videocassettes might foster affection and intimacy in the private enclave of the home. But those feelings always existed in contrast to the perils of public space. In public a videocassette represented the state's power over people's everyday lives. These two sensations were interconnected but distinct. By participating in this distribution system, Iranians came to distinguish between different standards in public and in private. Videos may have been a source of pleasure in private but they were not to be revealed in public spaces or institutions, which were controlled by the state.

The tensions between video's public peril and private pleasure are paramount in Babak Anvari's *Under the Shadow* (2016). The horror film, set during the early years of the Iran-Iraq War, chronicles life in a Tehran building as its residents contend with the realities of war. Constant power cuts, bombings and missile attacks, and nights spent together seeking shelter in the basement become part of everyday life for the small group of neighbors. Gradually each resident gives up on the building, fleeing the city for some place safer. When only one last resident, Shideh, and her daughter, Dorsa, remain, the scene is ripe for the perfect ghost story, as the horrors of war seem almost indistinguishable from the possibility of a haunting, supernatural force.

In the film, among explosions and screeching, chador-clad banshees, Jane Fonda appears in a neon leotard, tights, and sweatbands. Fonda's workout tapes, popular globally in the 1980s,[50] are something of a ritual for Shideh, perhaps an effort to stay sane as the world around her seems to collapse (figure 4.1a, b). As Shideh does her daily workout at one point in the film, there is a knock on the door. Paranoia runs high, as anyone could be on the other side, from a neighbor warning them to take cover to a djinn ready to steal their souls. Yet Shideh's first thought, before putting on her head scarf and cautiously opening the door, is to hide her Jane Fonda tapes and the video player. At another point, Shideh scolds Dorsa for leaving the tapes visible when someone comes to the door. The fear of getting caught with videocassettes is no less severe than the fear of being possessed by a mysterious demon. When Shideh opens the door to reveal a worker who has come to repair the shattered glass windows, we as viewers sigh with relief. But questions linger for the media scholar. How did home video technology end up in this ghost story? What kinds of horrors does it represent?

Under the Shadow was never formally released in Iran. Through informal means, however, it found viewership there, as it did all over the world. In

Figure 4.1a, b
(a) Jane Fonda's *Workout* became a way of life for many Iranians during the Iran-Iraq War. (b) In *Under the Shadow*, the *Workout* tape serves as a means of escape for Shideh but also introduces new problems, as the fear of getting caught is constantly present. *Source: Under the Shadow* (dir. Babak Anvari, 2016).

fact, the film came up in several of my interviews. Just as the Jane Fonda workout tapes resonated with many of my interlocutors, so did the way in which the film positions videocassettes at the threshold between public and private. As a British production, *Under the Shadow* had no reason to go to the censorship board in Iran and could therefore represent that threshold in unprecedented ways. In Iran, the screen is considered a public space, subject to the same modesty laws that govern a city park, shopping mall, or administrative building. Although religious law does not require that women wear the veil at home with immediate family, women characters must always be veiled in movies and television shows, even when they are among family and in the comfort of their own home.

When *Under the Shadow* shows Shideh putting on her veil to answer the door, it is a rare glimpse into what this movement between public and private looked like at the time. Such a scene could not appear in Iranian cinema, where a distinction between public and private does not exist on screen. It is significant that videocassettes also take part in this moment. They, too, were implicated in the transformation of a private space into a public space. For many of the people with whom I spoke, quickly hiding videocassettes was a familiar experience. Like Dorsa, they were also reprimanded for leaving videocassettes out where just anyone could see them. Above all, *Under the Shadow* reminded them of how scary and confusing the potential of getting caught was. The horror genre captured these complex emotions. This was a frightening time in Iran. A violent war devastated the country, while a new government asserted its authority through increasingly restrictive laws and policies. *Under the Shadow* shows how videocassettes were an important part of this moment in Iranian history—part of its terrors but also a mundane feature, often just playing in the background.

Videocassettes are a quick reference in *Under the Shadow*. Regardless, those Jane Fonda tapes create an opening for us to think about the role videocassettes played in everyday life during the 1980s and 1990s in Iran. What did it mean to grow up with videocassettes during these profound transformations? And how did young people's experiences with the circulation of videocassettes shape their understanding of public and private space? As my oral history interviews revealed, videocassettes helped untether the idea of crime from notions of morality for many Iranian children. Through their possession of Disney movies and classic Hollywood and old Iranian films, young Iranians came to appreciate that to commit a crime in the Islamic Republic did not always equate to doing something morally wrong. Embedded in that lesson was the idea that one always had to be careful to hide one's crime. Videocassettes were part of larger concerns of surveillance at the time and central to debates about who was responsible for a child's moral wellbeing. In this way, videocassettes did not change life in Iran per se but rather mediated what life meant under the new Islamic Republic.

The classroom was a space where Iranian children became acutely aware of the contradictions between the state and the home and between what was allowed in public and what was allowed in private. Videocassettes often featured heavily in these revelations. For example, Anaheed, who grew up in Tehran in the 1980s, remembered vividly the day that her mother took her to register for kindergarten: her pink dress, the building's tan walls, and how excited she felt that it was finally her turn to enroll in school. At six years old, she was the youngest of three children. As they walked, she and her mother talked about books and pens and school uniforms. Anaheed's excitement quickly turned into fear as they approached the school. Her mother suddenly stopped, pulled her to the side by the arm, and crouched down. With uncharacteristic sternness, she warned Anaheed that when they were in the school, she was under no circumstances to mention the fact that they had a video player at home. The command came completely out of the blue. Dreaming about her future classmates and the games they might play on the playground, videocassettes had been the furthest thing from Anaheed's mind. But after her mother's strict warning, all Anaheed could think about as they entered the school building was their video player, the small stack of children's movies hidden on the shelf behind the classics of Armenian literature, and the quiet man with a friendly smile who came with new videotapes every week. She was scared into silence, still confused about her mother's seriousness. As part of the registration process, a school administrator asked Anaheed a series of questions. When she asked, "Do you have videos at home?" Anaheed froze and tried not to look to her mother. She stuttered through a quiet no. She had been forced into her first lie. For Anaheed her first day of school will always be about videocassettes.

It wasn't until much later that Anaheed came to understand fully what had happened on that day. It was well known among mothers that administrators asked prospective students about videocassettes at home during the registration process. A rumor even circulated that if students answered in the affirmative, they wouldn't be allowed to register for school. Although the question was intended on the part of the state to regulate videocassettes, in practice it just kept them hidden underground. What was at stake in a memory like Anaheed's was children's education itself. The state tried to control video by threatening access to education. People may have learned to navigate around these maneuvers, but there was still a lesson for Anaheed in all of this. She felt a heavy sense of responsibility to protect her family's secrets. She learned that sometimes the pleasures of the home could not be discussed outside of it, and above all she came to see school as a place of lies as much as it was a place of knowledge. While in hindsight this experience made perfect sense to Anaheed, these were difficult lessons for a young child to grasp. What she could hold onto, however, were the videocassettes themselves. Her mother told her not to mention the video player at home, and she didn't. Her mother's stern disposition outside of the house made it clear that rules in public were different, and she respected that.

It is not a coincidence that videocassettes would feature so heavily in Anaheed's memory of this milestone, her first day of school. For an entire generation of Iranians, home video technology was part of education—not in the form of educational materials but rather as a foil through which they came to understand how right and wrong were not universal categories but rather conditionals whose meanings shifted as cassettes and people moved between spaces.

For some young Iranians, the charged rhetoric at school about video proved far more confusing and even challenged how they understood themselves as religious subjects. One man I interviewed, Mahmoud, who grew up in a family of film lovers, recounted a memory he had from elementary school. In the early 1990s, when he was six or seven, he arrived at school and went straight to his Quran class, a requirement outside of the normal academic units. He sat intrigued as the teacher began drawing a diagram on the chalkboard. On one side, the teacher drew heaven with puffy clouds and soft rays of sunshine, and on the other side, he sketched the lines of hell, where aggressive, jagged lines took the shape of fire. Mahmoud's curiosity increased as the teacher continued drawing. In the middle of his creation, the rectangular shape of a VCR began to appear. As he added the finishing touches, the teacher drew a line connecting heaven and hell straight through the VCR. This, the teacher announced, was the road to hell. Mahmoud was shocked.

He had spent many evenings watching Disney movies with his family on their old video player. He didn't want to spend eternity in hell but watching movies with his family was his favorite part of the day. He also didn't understand how videocassettes could be a bad thing, when they were part of his family's daily ritual. For the rest of day, he was distracted wondering if he and his family were destined for hell for their nightly movie-watching ritual. As soon as Mahmoud ran home, he asked his mom whether his teacher was right. Was their video player a direct conduit to hell? She assured him that using a VCR wasn't a sin but also warned him never to let anyone at school know that they had a video player at home. Mahmoud reiterated to me that his parents were both religious and prayed five times a day, so this incident did not represent a clash between religion and secularism. He described it to me as a watershed moment, as he learned to distinguish between the public and private spheres, which were often at odds with one another.

These stories from the schoolyard matter because they give us insight into how people, especially children, negotiated the state's harsh rhetoric about video. Through these negotiations they learned what it meant to be a citizen in the new Islamic Republic. As Roxanne Varzi has argued, schools played an important role in shaping "Islamic subjects" in the early years of the Islamic Republic.[51] She deploys the term "Islamic subjects" rather than "Muslim subjects" to denote the fact that the Islamic Republic insisted its citizens outwardly perform the state's interpretation of Shi'a Islam, regardless of their personal, private beliefs.[52] Perhaps unexpectedly, videocassettes helped many young Iranians understand exactly what Islamic subjectivity meant to their everyday lives. Mahmoud grew up in a religious family. He was Muslim, but that didn't necessarily mean that his family's religious values aligned with the state's. The video lesson in his Quran class taught him that the religion he learned at school—the state's religion—wasn't necessarily the same as the religion he practiced at home.

In time, my interlocutors internalized these unwritten rules, and they came to see their schools as state institutions at odds with home video technology. Arang, who was a teenager in the mid-1990s, recalled how one of his favorite pastimes was trading movies on videocassette with his friends from school. This was after the ban on video technology had been lifted, but that did not mean that all content was permissible. Years of keeping conversations about videocassettes private had taught the boys not to risk bringing them into the school building. A few mornings a week, the boys met at Arang's house, which happened to be around the corner from their high school, in order to exchange cassettes and talk about movies.

Memories like Anaheed's, Mahmoud's, and Arang's are hardly the exception, and in my interviews I heard about many such instances. Especially common were stories in which parents asked children to lie at school about the fact that they watched videocassettes at home. This was a big responsibility for a child but instructive in what it meant to live separate public and private lives in Iran in the 1980s and 1990s. As we can see, the state's rigid stance on home video technology often pitted parents against teachers, with children caught in the middle. This is not an entirely new phenomenon. Children have historically been at the center of the most pessimistic accounts of new media technologies.[53] From dime novels to television to video games, cynics have expressed concerns about what media content children consume and how often they consume it. Are children being exposed to too much violence? Will they become addicted to their devices? Certainly, these same concerns circulate alongside today's digital technologies.

For historians, however, the validity of such conflicting opinions is less important than the fact that they exist at all. In other words, new media technologies are often met with the dissenting voices that activate the rhetoric of a child's wellbeing. What we can do is situate each instance in the larger social and cultural context of the time in order to understand how these anxieties reflect bigger transformations at the time. In Iran in the 1980s and 1990s, children found themselves caught between their teachers and their parents, between their families and the state precisely because Iranian society itself was still figuring out the meaning of public and private under a new government and during war. To what extent would public and private be kept separate? And to what extent would public control extend into the private lives of citizens? These questions, anxieties, and uncertainties were embedded in the lessons and lies about home video technology.

THE HOME VIEWING NETWORK

"Any day except Monday," the video store owner in western Tehran told me as I tried to set up an interview with him. That was the day that *Shahrzad* was released every week, and he knew that his store would be swamped with customers crawling over each other to get their copy of the most recent episode. In April 2016, Tehran was experiencing what might be best described as *Shahrzad*-mania. The series *Shahrzad* had won over many Iranians, young and old alike. The show, which is now awaiting its fourth season, takes place in the 1950s. It tells the story of Shahrzad, a young medical student who finds herself caught in a love triangle as important political events unfold, including the CIA-led coup that overthrew the prime minister Mohammad Mosaddeq in 1953. Within just a matter of weeks of its debut in October 2015, the show found success. By the following April, as its first twenty-eight episodes ended, it had developed a cult-like following. The show was everywhere that spring. Posters adorned store windows at every turn, and coffee shops and cafes were abuzz with people speculating what Shahrzad's next move would be. Cheap knockoffs of Shahrzad's signature birdcage necklace were even sold in Tehran's Grand Bazaar alongside expensive yellow agate and turquoise rings. When I was in Tehran, Monday evenings were spent with friends and family watching the newest episode with bated breath.

Shahrzad was exceptional for several reasons. An Iranian series had never found so much success. It was made more spectacular by the stiff media competition of the mid-2010s, as Iranians had access to a seemingly unlimited supply of content from around the globe. The show also managed to represent love, gender, and politics in unprecedented ways. A strong female lead—who is a medical student in the 1950s, no less—captivates the men around her, and political dissidents are the heroes of the show.

These successes were not separate from the show's distribution. Unlike the serials and programs broadcast on Iran's public television stations and overseen by Islamic Republic of Iran Broadcasting (IRIB), episodes of *Shahrzad* were released as digital copies, both on DVD and online. Every video shop and corner store stocked DVDs of the most recent episodes of *Shahrzad*, and people also had the option of purchasing a copy online and downloading it. Because the DVDs were cheap, most people I spoke with preferred buying a disc of the episode each Monday, rather than wasting their limited internet package on such a big file. *Shahrzad* was distributed through a growing network called *Shabakeh-ye namāyesh-e khānegi* or the Home Viewing Network, which is a collective of supermarket suppliers, bookshops, and video stores that funnels certain DVDs into popular commercial spaces. *Shahrzad* wasn't the first show whose producers had made use of the Home Viewing Network's unique distribution methods. Comedic legend Mehran Modiri had also used the network to distribute his satiric comedy shows *Qahveh-ye talkh* (Bitter coffee, 2010) and *Shukhi kardam* (I'm just kidding, 2014). While Modiri's comedies were successful, with *Bitter Coffee* selling upward of fourteen million DVDs, they failed to achieve the same level of success and popular outpouring as *Shahrzad*.

Scholars and commentators agree that *Shahrzad*'s transgressive elements—the representation of women performing music on screen, for example—are the direct result of its distribution methods.[54] Like all cultural products in Iran, *Shahrzad* was carefully regulated by the Ministry of Culture and Islamic Guidance (MCIG). And yet, without a doubt, *Shahrzad* exceeded the limits of what official television broadcasts normally permitted, just like *Bitter Coffee* and

I'm Just Kidding before it. As a production specifically for home distribution, it crossed otherwise impermeable boundaries. In fact, the very emergence of the Home Video Network can be traced back to a disagreement between Modiri and Islamic Republic of Iran Broadcasting (IRIB) over *Bitter Coffee*. The Home Viewing Network offered producers a level of freedom that IRIB could not. As scholars try to understand the success of *Shahrzad*—whether the show's cutting-edge representation of women or its sharp critique of political history—the idea of digital distribution never lags far behind. The linkage between digital technologies and *Shahrzad*'s taboo-breaking representations is made especially clear in the title of Khatereh Sheibani's recent article "*Shahrzad*: Pop Culture, Media Culture, and the Representation of Gender in the Digital Age." And yet in all these accounts of *Shahrzad*'s success and the Home Viewing Network, the unique underground video infrastructure of the 1980s and 1990s remains strangely absent.

The cult-like success of *Shahrzad* cannot be boiled down to a single reason. The analysis of the show's content, especially its take on gender and politics, goes a long way in explaining why the show has fared so well over three seasons. While scholars like Sheibani have argued that the new digital technologies that support the Home Viewing Network have "enticed viewers,"[55] I would like to make the opposite claim. Far from appearing new and exciting, the Home Viewing Network is old and familiar. It is a continuation of the underground video rental network from the 1980s and 1990s. It was in the era of videocassettes that Iranians first rejected IRIB and other state-controlled media. That is when they first understood home media consumption as a choice, and one with potential political power.

For many of the people with whom I spoke, watching *Shahrzad* every week also represented a return to family viewings. As Leila, a Tehran-based *Shahrzad* fan, explained to me, "One of the reasons I am so attached to *Shahrzad* is because we have turned watching it into a family ritual. In the 1990s, we used to watch movies on video together all the time, but somehow that ritual faded away. Now with *Shahrzad* it's like old times. My mom, dad, brother, and I even sit it in the same spots we used to sit in when we would watch American movies together twenty years ago." *Shahrzad* evokes feelings of nostalgia not only because of its historical narrative but also because its home exhibition reminds viewers of past media practices.

The Home Viewing Network that supports the distribution of shows like *Shahrzad* provides an infrastructure for "officially vetted productions intended to be watched within the private sphere of the home."[56] While this is certainly the case now, the idea of the home as a space that mediates public and private interests is not new to Iran. Decades before *Shahrzad* and the Home Viewing Network found success, the underground video distribution network captured the tensions between public and private interests. Videocassettes moved secretly between and through public and private spaces and as a result materialized in the social and political climate at the time.

While the state may have demonized video technology—not unlike the ghost that haunts *Under the Shadow*—everyday users' relationship to the medium wasn't so simple. For the most part, people's attempts to remember videocassette technology were marked by fondness. These were the pleasures of videotapes—an opportunity to experience the world through the small screen; the victory of watching a movie with distorted sounds and images; and the intimacies fostered with and by media technologies. Lurking in the corner, however, were the perils—the fear of being caught; the responsibility put on children to protect their families' secrets; the uncertain realities of rejecting state control; and the confusing pull between public and private spheres.

This chapter has shown how home video technology was not separate from the home. Videocassettes were a domestic technology that helped forge the home as a unique space in post-revolution Iran. As the home opened to more and more functions, people experienced videocassettes in every manner possible. In one moment, a cassette was the world; in the next, a whisper in the ear. For an entire generation, videocassettes were indistinguishable from the experience of growing up as a young citizen in the Islamic Republic of Iran.

VIDEO MATTERS: REMEMBERING THE UNDERGROUND

During a particularly productive interview with a middle-aged couple, Fariba and Bahram, and their two children, both in their early twenties, Fariba stood up, clasped her hands, and said enthusiastically, "Do you want to see our collection? We still have our cassettes!" There, in the fourth-floor apartment near Vanak Square in Tehran, the entire family led me to the eldest daughter's room. When I peeked in, I expected to find a small shelf or a TV stand that held a few cassettes. Instead, there weren't any signs of video at all. The only discernable storage—a bookshelf facing the twin bed—housed piles of novels and textbooks, but I didn't see even one videotape in the mix.

Then, all at once, every member of the family descended on a different part of the room: under the bed, in the wardrobe, behind the laundry basket. They pulled out box after box and opened them with delight. Each one was packed with videocassettes. The family showed me tapes with home videos, music videos, Turkish TV shows taped off of satellite TV, and their favorite movies. Some movies were in original sleeves, while others were clearly copies—a familiar generic VHS cover with "*Pari-ye daryāi-ye kuchak*" (*The Little Mermaid*) scribbled on a label on the front of the cassette, for example. But as the family members clambered to show me their favorites or to tell me a story about this cassette or that one, *original* or *copy* didn't matter. Both carried the same emotional weight.

In the span of half an hour, the daughter's room had transformed into a cluttered video archive. Cassettes sprawled out on every surface, filling the room with plastic, electronic ribbon, memories, and emotions. "*Yādesh beh kheyr*," Fariba finally sighed, invoking a common phrase used by Persian speakers when fondly remembering the past, something akin to "good times" or the "good old days." The others nodded their heads in agreement and echoed, "*yādesh beh kheyr*," before packing up the cassettes and returning them to their hiding spots, where they would continue to gather dust.

Years after the family had disconnected their VCR, these cassettes lived their lives out in waiting—for moments like this when they might be held again but never watched.

My experience in that apartment near Vanak Square, though special, was not unique. While most of the interviews I conducted took place in public spaces, like cafes and restaurants, the people I interviewed, both regular consumers and dealers, were always eager to tell me about the collection of videocassettes they kept tucked away. Often, they would send me pictures of their video collection as soon as they got home. Online platforms that allow users to sell second-hand items, websites like *Divar, Esam*, and *Sheypoor*, still regularly feature ads for videocassettes, both blank cassettes and those with prerecorded content. While careful to note that all of the movies are *mojāz* (permitted), they also rarely list specific film titles. Instead, customers purchase a grab-bag package without knowing exactly what they will get. All of this is to say that videocassettes are still very present in Iran—big, bulky reminders of a pastime that no longer exists the way it once did.

I wondered why people continue to cherish these small personal video archives long after analog home video technology has fallen out of favor. Here I was less interested in the specific media content that individuals have held onto and more interested in the kind of technology that they have chosen to preserve. The middle-class Iranians I spoke to were not media collectors, hoarders, or technology enthusiasts by any means. They did not keep everything indiscriminately or possess a particular interest in collecting technological artifacts. The apartment lifestyle that defines contemporary urban life in Iran wouldn't allow it. My experience at Fariba's apartment in Tehran showed how limited space is in the city.

By comparing analog video technology to other media technologies, it becomes clear that videocassettes occupy a special place in people's hearts and in their homes. In Iran, digital technologies overtook analog systems as the primary mode of viewing prerecorded content in the late 1990s, first through VCDs, then DVDs, and more recently video on demand (VOD). DVDs remain in active circulation in Iran, both sanctioned and not. They are cheap and the selection is vast. Yet the people with whom I spoke did *not* keep the DVDs that they purchased, with the exception of a few of their favorites. If, as Jennifer Gabrys suggests, "The dump is a site where objects typically absent of utility or value collect,"[1] then what can be said about the videocassettes that have, to some extent, been spared the dump? What value do they still hold for Iranians? What memories do they represent? Such questions respond to Charles Acland's call to pay more attention to "residual media," or the remains of old media that continue to fascinate users today.[2]

In this chapter, I explore how, since 2010, people have sought to remember, record, and represent the unique history of analog video technology in Iran. I examine a wide range of materials, including novels, magazines, movies, and social media platforms, in order to reconstruct the practices and debates that have continued to form around videocassettes in Iran. While the previous chapters have explored oral history as a site of remembering, in this chapter I focus on other discursive forms that might instruct us on the ways in which video has continued to be constituted and negotiated.

Since 2015, Iranians have become increasingly invested in remembering the videocassette era. Like their counterparts elsewhere in the world, they have mourned certain global technological losses, like the end of VCR production in 2016.[3] But more often, they have engaged with the particularities of the underground video infrastructure in Iran, including the material traces it left behind. These contemporary reflections establish video not just as a media technology but also as a unique cultural practice with its own history, emotional drives, motivations, processes, and forms of engagement.

By focusing on the ways in which Iranians have remembered and codified the experience of videocassettes since the turn of the century, I argue that analog video technology continues to shape the meanings Iranians assign to media. The videocassettes still present in many middle-class Iranian homes no longer represent entertainment or communication but rather are emblematic of the cultural and social structures and values that determined how they ended up there in the first place. In short, movies on videocassettes still *matter*. Marianne Hirsch and Leo Spitzer have claimed that material remains can serve as "testimonial objects" that allow "us to focus crucial questions both about the past itself and about how the past comes down to us in the present."[4] In tandem, I've shown throughout this book that the historical junction at which analog video technology emerged in Iran, a period of political instability and wartime violence, coincided with the materiality of this new medium. This collusion of forces turned the business of circulating and consuming media into an underground practice, where it has largely remained. It is no coincidence then that those same videocassettes that once activated, illuminated, and animated networks of media distribution now await chemical decay in the same underground spaces from which they originally emerged.

By collecting and archiving videocassettes, Iranians have continued to relate to the underground video infrastructure. At the same time, new representations of past video culture also show how video still matters. By examining popular movies and social media accounts, I reflect on the ways in which the past, present, and future of media intertwine. As Wendy Hui Kyong Chun cautions, the discursive future that new media construct is specifically grounded in "data of the past."[5] If we look to entertainment and social media, we also see a preoccupation with the past, even as digital technologies promise to transform the future of communication. From Instagram accounts dedicated to the forgotten years of our childhood to TV reboots on streaming platforms like Netflix, digital media seem to be just as entrenched in the past as they are in some imaginary future.

In Iran, too, contemporary media intersect with a collective hankering for the past. These engagements with yesteryear often form around technologies that once shaped entire generations of media consumers but have since fallen out of favor. Of special concern here are the recent popular films and social media accounts that have revisited the history of analog video technology, opening up spaces for people to reflect on this underground infrastructure and the affects and communities that formed around it. Such efforts construct video as a specifically Iranian practice and emphasize the historical, political, and material circumstances that were unique to videocassettes in Iran. They also show how contemporary media practices are still very much bound up with the past.

In some respects, the larger questions this chapter grapples with apply beyond the borders of Iran. Of most national contexts we can ask, where did all of the videocassettes go? How have people abandoned or tried to preserve the practices that developed around the material objects themselves? What does the afterlife of video tell us about the values and meanings that people continue to ascribe to media, whether emergent, steadfast, or in decline? Jeff Scheible, drawing primarily on American and European sources, has shown that "just because physical video stores 'close,' their material and cultural traces do not simply vanish."[6] In order to keep track of the afterlife of videos as a new material order, Scheible theorizes what he calls *redistribution* to describe the ways in which media objects continue to circulate long after we expect them to. Accordingly, redistribution comprises the processes that allow for the "movement of media objects that have already participated in cycles of distribution and reception."[7] In this configuration, redistribution refashions and extends the circulation of media objects. This productive take on media distribution applies to Iran, too: videos continue to zigzag throughout the country, not just materially but also discursively in the form of values and practices.

PUBLIC MEMORIES OF THE UNDERGROUND

The videocassettes that once sustained the video infrastructure have disappeared from sight—tucked away in boxes, under beds, and in basement storage rooms. And yet over the last several years, memories of underground video culture have become highly visible. Popular media, including cinema and digital platforms like Instagram, have functioned as public sites where Iranians negotiate, remember, and articulate their recollections of the underground. If the video infrastructure of the 1980s and 1990s was a public secret that people engaged in without acknowledging it, then remembering it several decades later has become a public pastime. Once banished to hushed whispers behind closed doors, home video technology and the illegal world it inspired are now shown on screen and talked about openly. As I will show, popular representations of video culture publicize the underground infrastructure and its history, thus normalizing informal and illegal media systems.

A string of recent popular movies has remembered underground video culture from the 1980s and 1990s. These films are remarkable not only because of their representation of the history of video decades after the analog technology fell out of favor but also because of their success at the box office. Saman Moghadam's *Nahang-e 'anbar* (Sperm whale, 2015) and its sequel *Nahang-e 'anbar-e 2: selekshen-e Roya* (Sperm whale 2: Roya's selection, 2017) exemplify this trend. The *Sperm Whale* movies, which broke ticket sale records, do not focus exclusively on underground video culture but rather reflect on growing up during the revolution and war. Nevertheless, the movies go to great lengths to reconstruct videocassette circulation and thus make visible the centrality of the medium to everyday life in the 1980s. The *Sperm Whale* franchise offers a nostalgic look back at video practices, while foregrounding the medium's illegality. The films' comedy pivots on exposing the public secret of underground video culture. They make parts of the video infrastructure visible and poke fun at the policies and processes that determined how videocassettes could circulate at the time.

The *Sperm Whale* movies write the underground video infrastructure into the official history of Iran through a series of flashbacks. Their tragic hero, Arzhang Sanubar—who is precisely of that generation who came of age in the late 1970s and early 1980s—navigates viewers through the most transformative events of the Islamic Republic's first two decades, including the revolution and the war. And yet video is never far off. As Arzhang states early in *Sperm Whale 2*, "I don't know what you guys did at the time, but we were always watching movies." Of course, here, he means movies on video, a medium that viewers see often in both films. In fact, Arzhang spends the better part of his youth as an underground video dealer, even introducing himself to one of his customers as Sanubar the Filmi. While other Iranian movies and television shows from the last decade have shown home video technology on screen, it has either been relegated to the background or defined as part of a criminal underworld. In contrast, the *Sperm Whale* movies make video a main character and thus show the important role that it played in people's lives during the Islamic Republic's first several decades.

The first *Sperm Whale* movie gives viewers unfettered access to underground video distribution in the 1980s. Scrapped for cash and wanting to impress Roya, his long-time love interest, Arzhang finds himself caught up in the business of video rental. *Sperm Whale* portrays his work as a filmi in astonishing detail. The viewer sees him walking through the streets with his briefcase full of cassettes, his personal video archive. The viewer even experiences the fear of getting caught vicariously when someone unexpectedly walks in on Arzhang and a customer (figure 5.1a, b). Slow motion, sunglasses, and carefully placed music highlight Arzhang's new role as a filmi but also position this kind of work as venerable and important. Arzhang's makeover as he transitions from an awkward teen to a stylish filmi is an important transformation in the film. It marks the beginning of what will ultimately become a successful adulthood, albeit not one without tribulations.

Full of bright colors and plenty of humor, the scenes of video distribution in *Sperm Whale* are unlike most representations that came before it. These scenes are not burdened by a moral message. Instead, Reza Attaran, the famed comedic actor who plays Arzhang, turns the underground circulation of videocassettes into an inside joke—as he often does in his work.[8] These moments in the film are funny not just because Attaran,

Figure 5.1a, b
(a) The process of browsing videocassettes and (b) the fear of getting caught are
depicted in *Sperm Whale*.
Source: Nahang-e ʿanbar (dir. Saman Moghadam, 2015).

a middle-aged actor, plays the twenty-year-old version of his character in
flashbacks, but also because the filmi is a figure that most people remember
from this period and yet have rarely seen represented in this way, especially
in popular media (figure 5.2a).

Although the underground video infrastructure may exist without an
explicit moral message in *Sperm Whale*, it does not leave the law behind. The
medium's illegality continues to haunt its representations even in the most

humorous scenes. At one point in the film, Arzhang decides to sell his video business. As he is transporting all his video equipment to its new owner, police discover his huge bundle of contraband videocassettes and players clumsily strapped to the top of his car. They pull him over and arrest him. At the very moment he tries to abandon video, Arzhang becomes victim to its illegality. In a series of comical exchanges, Arzhang tries to defend video to the police by claiming that he records religious sermons and shares them with people who cannot otherwise see them. Despite his objections, he is dragged into the police station, where his collection—full of unpermitted films—reveals his true motivations. Arzhang's fate does not look promising, but he is ultimately saved by an acquaintance from his time on the war front, a powerful cleric who vouches for Arzhang and then employs him as a bodyguard.

Sperm Whale manages to represent this controversial period in astonishing detail, including police station shots that perfectly frame Ayatollah Khomeini's photograph (figure 5.2b).[9] As Babak Tabarraee has suggested, Reza Attaran's unique status as a veteran deployed not just once but three times during the Iran-Iraq War has afforded him certain privileges when it comes to representing governmental restrictions, especially during this war period.[10] Analog video technology and its ban play a central role in Sperm Whale. The humor in many of these scenes hinges on the absurdity of a video ban in the first place. From underground dealings to police interrogations, these scenes are as tragic as they are hilarious.

While Sperm Whale details how videocassettes circulated as material objects and the laws and policies that structured their movement, Sperm Whale 2: Roya's Selection reveals what people were watching on video and how. In Sperm Whale 2 videocassettes do not just appear as nameless contraband items that traverse city streets unseen; they also house audiovisual content that means a lot to their viewers. The critical reception of Sperm Whale 2 was less favorable than the first, with critics accusing director Saman Moghadam of depending too much on the jokes and antics that made the first movie popular, rather than developing a new robust storyline.[11] Nevertheless, Sperm Whale 2 was an enormous hit at the box office, becoming the best-selling film in Iran's history.[12] The success of Sperm Whale 2 shows how its representation of movie culture in the 1980s resonated with viewers. In it, we see the same kinds of shenanigans centered on the same kinds of stories: video dealing, arrest scenes, and plenty of music. The film is about life during the war, and the movies, friendships, and trips to the

Figure 5.2a, b
(a) Arzhang looking cool as a filmi and (b) a detailed police station scene are depicted in *Sperm Whale*.
Source: Nahang-e ʿanbar (dir. Saman Moghadam, 2016).

Caspian Sea that offered reprieve from the violence unfolding around the film's characters. In all these moments, videocassettes are present, not just as Arzhang's livelihood as a video dealer but also as entertainment and escape for the main characters.

In addition to scenes in which characters watch and debate movies, *Sperm Whale 2* renders videocassette content from that period onto the big screen. At one point, the characters recreate the music video for Michael

Figure 5.3
Reza Attaran takes on the role of Danny Zuko in an outrageous reenactment of
"Summer Nights" from *Grease*.
Source: Nahang-e ʿanbar-e 2: selekshen-e Roya (dir. Saman Moghadam, 2017).

Jackson's "Thriller." At another, they reenact the "Summer Nights" scene
from *Grease* (dir. Randal Kleiser, 1978), with Attaran in the role of John
Travolta's character, Danny Zuko (figure 5.3). These scenes, which proved
popular with both audiences and critics, are particularly important to what
Sperm Whale 2 achieves in its memorialization of this period. Michael Jackson came up often in the interviews I conducted, because his music videos
filled the remaining tape in many of the videocassettes that users would
rent. Urban Iranians developed enthusiasm for Jackson's music through
these clips at the end of movies like *Grease*. As we have seen, these experiences had previously been limited to poor-quality copies on the small
screen at home. *Sperm Whale 2* takes this ephemeral content and renders it
in high definition on the big screen. In doing so, it restores a lost history.
Given the controversy surrounding pop music in Iran in the early years
after the revolution,[13] global touchpoints and citations like Michael Jackson
are absent from Iranian productions in the 1980s.

Sperm Whale 2 shows what people in the 1980s were actually watching,
while also inviting traces of a forbidden media technology into a public
space like the movie theater. As one interviewee told me when I mentioned
that I wouldn't be able to see *Sperm Whale 2* in theaters: "You have to watch
it in the movie theater! Michael Jackson's 'Thriller' right there with all these
people—it's unlike anything you've ever seen before!" The scene on the big

screen was irresistible to many Iranians. Shortly after the movie's release, platforms like YouTube, Aparat, and Namasha were flooded with videos of the "Thriller" and "Summer Nights" scenes, which viewers had shot on their phones in the theater. In each case, the audience's laughter drowns out the movie's soundtrack. *Sperm Whale 2* takes old video content, jazzes it up, and turns it into a new public experience. Taken together, the two *Sperm Whale* movies construct a distinctly Iranian narrative of video: a wartime technology whose materiality mattered. The franchise's appeal to certain emotions like nostalgia and the humor—which made it successful—depends on the triumphs and failures of video technology in Iran in the 1980s and 1990s.

Scenes in the *Sperm Whale* movies that detail the logistics and pleasures of underground video distribution are significant because they render visible a world that now exists primarily in people's memories. Prohibitions on video technology and the nature of state control in Iran meant that very few representations of video culture existed during the 1980s and 1990s. As we saw in chapter 3, contemporary sources—films like *Song of Tehran* (1992), *Soltān* (Sultan, dir. Massoud Kimiai, 1997), and *Dealer* (2001)—portray video primarily in terms of its criminality, often equating video distribution with dark spaces, criminal life, and poor morals. *Sperm Whale* turns those representations on their head by portraying video culture as normal and celebrated, even as it exists on the margins of legality.

In fact, it is the laws forbidding videocassettes that are actually absurd in these representations. The video dealer becomes the film's hero, while the state becomes the butt of many of the film's jokes. The *Sperm Whale* franchise thus captures the contradictions of the underground, as video is illegal but normal; banned but everywhere; and hidden but just out of sight. That many of my interlocutors referenced *Sperm Whale* and *Sperm Whale 2* to describe their own experiences with video speaks to the movies' resonance for viewers. The representation of video is a touchstone through which viewers access memories of a very different time and place, when movies were on cassette and video dealers quietly ruled the streets.

Beyond film, other media have become spaces that make public the history of underground video. Instagram, in particular, serves as a site of remembrance for videocassettes in the late 2010s. In Iran, as elsewhere, Instagram is not just a platform for sharing pictures. It is a marketplace, where people advertise and sell goods, a dating app, where youngsters meet and flirt, and so much more. Because, unlike Facebook, Instagram has not been filtered by the government, young urban Iranians are particularly active on this platform and spend countless hours posting, sharing, and chatting. They bend and exceed the intended uses of the application. It is a site where people post about their present—about their lives today—but it is also a place where people remember the past. Recently, Instagram accounts from Iran have been speckled with familiar sights: stacks of videocassettes and piles of analog video equipment like VCRs. These pictures come with hashtags like "nostalgia" or "good times" (*yādesh beh kheyr*). These posts document how a certain generation of Iranians continues to cope with the material and affective residue of a fallen medium.

Recently, for example, an Instagram page dedicated to *shasti-hā*—a designation for those who were born during the 1360s on the Iranian calendar (approximately the 1980s on the Gregorian calendar)—posted a close-up shot of several old movies on videocassette, including a selection of Indian films like *Aulad* (Child, dir. Vijay Sadanah, 1987) and *Nasib* (Destiny, dir. Manmohan Desai, 1981), and a cassette labelled "Turkish Shows." The picture accompanied a caption that read, "Hello, can anyone clean the head of this video? Do you have rubbing alcohol and a cotton ball? Should we sit and watch a movie? These are some of the sentences that came with watching a movie on videocassette." This description ended in a long list of hashtags that include "childhood" (*kudaki*), "don't you remember" (*yādetun nemiyād*), "nostalgia" (*nostālzhi*), "our childhood" (*bachchegimun*), "memories" (*khāterāt*), and "the generation of censorship" (*nasl-e sānsur*). The post received nearly a thousand likes and generated comments that ranged from nostalgic utterances like "*yādesh beh kheyr*" to more complicated ways of remembering that emphasize the technological or political context of these cassettes. One user wrote, "We enjoyed those films with poor picture quality even more. Good times!" Another user jokingly recalled "I remember one time we hid all of our videocassettes in the A/C duct. Haha." Others, though, took a more cynical view of this period, "Good times, when the punishment was death." Another person similarly remembered, "In those days, they were considered contraband. If they ever caught anyone, the punishment was death. At least a hundred times we gathered everything up but then would secretly bring them out again."

In these acts of remembering, the format of video mattered more than any particular content. Certainly, some commenters referred to Indian movies in general terms, noting the "archive" of Indian films they had or the fact that they would "watch an Indian movie every day." They nevertheless paid very little attention to the content of the actual movies displayed in the post, even though the picture focuses mostly on their titles.

Instead, users remembered everything else involved in watching a movie on video, from cleaning its head to negotiating its legibility to accessing and protecting it from the law. Another post from the same account echoed many of these same sentiments:

> Hello, good morning. Do you remember when #video #video_cassettes were illegal? When we wanted to move them, it was as though we were mobsters doing something horrible. Because of these [cassettes], they scared us so much. They stole so many cassettes, but what was the reason? Just like so many other stories. Do you remember waiting for the music videos at the end? They had tracking. Remember having to take the ribbon out to clean it? Ooh man . . .

Posts like this remind us how memory and materiality intersect, while the comments they generate pull us back to a history of smuggling. Seven out of eleven comments on the post refer to transporting videocassettes, with memories of wrapping cassettes in newspaper or hiding them under clothes, or declarations such as "It was exactly like smuggling" ('ayn-e qāchāq bud). One commenter wrote, "What do you mean scared us? Because of these, I was arrested and beaten. The way they interacted with me, I felt like I had smuggled arms." Another person wrote, "Transporting two kilos of opium was less frightening than video," while a third user commented, "The way we had to move these, it was like drugs. But good times. We were happy with just that."

Even decades later, people still construct a narrative of video using the language that was intended to regulate it in the first place. In the 1980s and 1990s, the state invoked the moral degradation of things like arms smuggling and drugs to discourage the use of video. Iranians came to appreciate the movement of these goods in similar terms. But the irony in all these comments is clear: smuggling videocassettes is not the same moral violation as smuggling drugs or guns. The regular movement of videocassettes through people's homes normalized them and redeemed them from the cultural value of other contraband items. This can be seen in the punctuating comments like "Good times. We were happy with just that." Indeed, the fact that comments about arrests and beatings can occur alongside comments like "Wow, good times!" or "We would also wind audio cassettes with a pen or pencil" shows how diverse and dynamic the experience of remembering video has become.

Given the diversity of experiences and forms of remembering, we might ask why people have chosen to memorialize video now and in these ways. Popular representations have brought the underground video infrastructure above ground and into the light. In doing so, they have taken the private, hurried experience of accessing and watching a video and refashioned it into a public act of remembering. Just as many Instagram posts recount memories of home video technology through the language of longing (with expressions like *yādesh beh kheyr*), the critical reception of *Sperm Whale* and *Sperm Whale 2* has largely focused on the issue of nostalgia.[14] Popular representations of underground video are undoubtedly nostalgic engagements. The interest in nostalgia is hardly unique to Iran or to the present day. Nostalgic engagements with the past have long been a preoccupation of popular media, and scholars have been writing about the phenomenon since at least the 1970s.[15] As Svetlana Boym claims, "Nostalgia is a disease of the modern age."[16] As such, it is both pervasive but also rooted in conditions of a particular time and place. The kinds of nostalgic representations of underground video in Iran may be part of a global trend, but they are also tied to a very specific history of video in the country.

It makes good sense that nostalgia would be a dominant mode for representing the history of the underground. Nostalgia and the underground have much in common. The underground was structured through binaries such as public and private, formal and informal, and legal and illegal. Similarly, nostalgia is a feeling where past and present collide, and individual and group remembering take place together.[17] Nostalgia tells us just as much about the present as the past and about collective memory as much as personal feelings.

We might position instances of remembering underground video culture as examples of "media nostalgia" or a longing for fallen media as a way of coping with current technological and cultural change.[18] As the underground video infrastructure demonstrates, media technologies can never separate entirely from politics. Engagements with videocassettes on social media platforms and in popular cinema are nostalgic not because current media practices are *radically* different but rather because they are *slightly* different. Although analog media technologies have given way to digital platforms, Iranians still access and consume media primarily through informal means that operate outside of what the state has sanctioned. Thus, in these representations, the underground continues to be an infrastructure upon which Iranians enact a range of subjectivities, many of which still run contrary to what the state has envisioned for them.

OLD CASSETTES/NEW CINEMA

Throughout the 1980s and 1990s, the video infrastructure offered a safe retreat for movie culture at a time when the official film industry churned out state propaganda and movie theaters were physically dangerous places to be. In addition to a selection of world cinema, video provided access to the classics of Iranian cinema, including the banned midcentury commercial movies collectively known as *filmfārsi*. The underground infrastructure thus nurtured new generations of viewers for these movies since they were no longer available through formal means after the revolution. Recently, the effects of that incubation have started to bear fruit. A new popular cinema has emerged in Iran as filmmakers who grew up on video have begun to integrate filmfārsi conventions and tropes into their movies. These new comedies have found tremendous success at the box office with new records set almost every year. Curiously, a number of these popular films also meditate on the history of videocassette technology in the country. In addition to the *Sperm Whale* franchise, movies like Mostafa Kiayee's *Zedd-e goluleh* (Bulletproof, 2011) and Mani Haghighi's *50 kilu ālbālu* (50 kilos of sour cherries, 2015) represent the history of home video technology and are also part of this new popular cinema.

Given these examples, we might probe more deeply into why underground video culture is a central concern of these popular films. In other words, why have members of the commercial film industry—especially directors committed to popular filmmaking—chosen to remember video culture from the 1980s and 1990s? The critical reception of this new popular cinema often likens it to filmfārsi, the enormously popular commercial cinema of the 1960s and 1970s mentioned earlier. Although the formal production of filmfārsi began to die out with the establishment of the Islamic Republic, as we have seen, viewership of these films continued and even flourished through the underground video infrastructure. Movies like *Sperm*

Whale, *Bulletproof*, and *50 Kilos of Sour Cherries* reproduce the aesthetics and commercial values of filmfārsi, while also memorializing the video culture that once sustained movie culture in Iran.

During an extended stay in Tehran in the summer of 2015, I tried to see *Sperm Whale* three times before I was finally able to secure tickets. Each time it was sold out, and without an Iranian bank card to buy tickets online, I couldn't compete with the thousands of other viewers who wanted to see the comedy. I finally ended up at a matinee at the Charsu Cineplex in central Tehran, where I sat in a surprisingly crowded theater. I had seen the movie advertised all over the city, and it was a mainstay at most of the movie theaters that I had passed in my walks around Tehran. But it wasn't until the people I was interviewing started referencing *Sperm Whale* as a way of articulating their own experiences of video that I knew that I had to see it. As I expressed my frustrations during one interview at not being able to get tickets, the young man I was interviewing, a nineteen-year-old from northwest Tehran, asked, "What do you expect? Going to the movie theaters is one of the few pastimes that still exist for young people in Tehran." His comment confirmed my suspicions that, by 2015, movies as a public event for entertainment and leisure had become important again. This was a radical transformation from just three decades earlier when underground videos sustained a viewership that wasn't interested in going to movie houses anymore.

The 2010s gave rise to a new kind of popular cinema that viewers and critics were quick to liken to the filmfārsi tradition.[19] As film scholar Babak Tabarraee argues, this new cinema was driven largely by a surge in absurdist comedies. Long gone was a film industry fueled entirely by social realist films and slapstick comedies. Instead, a new outrageous style of popular movies emerged that was actually well-liked among young viewers. These movies have accounted for millions of dollars in ticket sales over the last decade.[20] For many, the tremendous success of these films was reminiscent of the popular cinema that had once populated movie theaters in Iran. One reviewer, for example, compared the success of *Sperm Whale* to that of *Ganj-e Qarun* (Qarun's treasure, dir. Siamak Yasemi, 1965) and *Qeysar* (dir. Massoud Kimiai, 1969).[21]

These comparisons mean a lot to the history of Iranian cinema. *Qarun's Treasure* and *Qeysar* proved the viability of the Iranian film industry, while also setting certain standards with respect to themes, aesthetics, production

values, and industrial practices. *Qarun's Treasure* was the first Iranian production to beat out a foreign movie in box office sales, dethroning Raj
Kapoor's *Sangam* (1964) as the highest-grossing film at the time.[22] It also
initiated certain conventions that would shape film production for decades
to come. By drawing these comparisons, some critics elevated the importance of *Sperm Whale* to that of *Qarun's Treasure* and *Qeysar*. They canon-
ized it, making it a turning point in the history of Iranian cinema.

It was not just the success of the *Sperm Whale* franchise and others in
the same vein that reminded viewers of the filmfārsi tradition. Their narrative structures, star power, and production values also shared similarities
with midcentury commercial cinema. According to Tabarraee, the "absurd
wave" in contemporary Iranian cinema prioritizes the audience's momen-
tary enjoyment over a careful narrative structure. While these films often
represent social issues, they do not take them seriously enough to provide
any meaningful commentary.[23] This trend's "brand of absurdity maintains
that absolutely nothing is serious."[24] For some reviewers, these qualities
recalled the filmfārsi tradition of years prior—and not in a good way. Many
critics consider filmfārsi a cinema of excess, both in the bourgeois values it
propagated but also in its extraneous characters, unnecessary plot twists,
and repetitive narratives. As one reviewer claimed, "*Sperm Whale 2* has all
of the makings of a modern-day filmfārsi," pointing out the film's weak
narrative structure and claiming it as derivative of Hollywood films like
Footloose.[25] Another reviewer wrote that "like all filmfārsis, *Sperm Whale*
contains sudden shifts in emotion, unexplained and superficial excitement,
and an illogical and unstable narrative structure."[26]

Several movies within this new popular trend include vivid representations of underground video culture, including examples by the two filmmakers most closely associated with the movement: Reza Attaran and
Mani Haghighi. Tabarraee sees similarity between the ways in which the
popular films of the mid-twentieth century inscribed bourgeois values that
served the shah's regime and the way in which Attaran's films service apo-
litical, "ambivalent values" in the years following Mahmoud Ahmadinejad's reelection in 2009.[27] These political arguments make sense. But recent
popular cinema is also connected to filmfārsi *infrastructurally* through videocassette technology. Popular movies like *Sperm Whale*, *Sperm Whale 2*, and
50 Kilos of Sour Cherries direct us to the circuits that kept the memory of Iranian cinema alive, even after public screenings of filmfārsi ceased to exist.

As midcentury popular cinema retreated from public movie screens in the early 1980s—to be replaced by a new kind of cinema sanctioned by the young republic—it found refuge in the circuits of video distribution that crisscrossed cities around the country. As chapter 1 showed, filmfārsi circulated widely underground on video in the 1980s and 1990s despite a ban on certain prerevolutionary movies. Reports from the time emphasize that, throughout the 1980s, Iranian films produced before the revolution were the most dangerous kind of film for a video dealer to distribute.[28] This was a sentiment that was also echoed in my interviews. One dealer, for example, insisted that he was never too worried about getting caught in the early 1980s because he did not carry the Iranian films that made video dealers vulnerable at the time. The fact that the state targeted popular midcentury commercial movies reveals that a demand for them existed.

That demand continued for generations. All of my interlocutors who were born after the revolution remembered watching filmfārsi movies for the first time on video. Often, these were some of their most memorable childhood experiences. The underground video infrastructure thus supplied them with access to a history of Iranian cinema that had otherwise been out of reach. The video distribution system nurtured a devotion to Iranian cinema for many of my interlocutors and encouraged a new generation of cinephiles.

Even filmmakers have acknowledged the link between the video culture of the 1980s and the success of Iranian cinema in the twenty-first century. Although the 1980s were a time of tumultuous political and cultural change and a period of violent instability, videocassettes helped ensure that people were never far from the movies, even if the nearest theater was many miles away. We can find evidence of this in a recent book-length interview with Asghar Farhadi, one of Iran's most famous filmmakers. In conversations with fellow filmmaker Esmaeil Mihandoost, Farhadi emphasizes the significance of videocassettes to his film education. The beginning of the interview covers Farhadi's childhood in a small town outside of Esfahan and his earliest encounters with the movies. "It wasn't easy for me to go to the movie theater," he explains, "because the only theater was in Esfahan and in order to watch a film, we'd have to make the trip there." That distance, however, did not keep him from watching movies at home. He continues, "At that time, Betamax videos had just become popular in Iran, and they were banned, too. We were one of the first families to have video.

I even remember one night my dad came home with ten or fifteen movies, and our task over the next two weeks was to watch them all."[29]

By responding in this way, Farhadi underscores the centrality of video-cassettes to his early interest in cinema. At the same time, the ban hovers in the background of his answer. In this interview, which was published in Iran in Persian, there was no need to clarify that videocassettes were banned. That information would have been common knowledge to most readers. And yet it shows how—as much as the ban was incidental because it did not actually constrict people's access to movies on video—it still structured how people understood movie culture at that time. During the interview, Farhadi is quick to emphasize that his film education on video did not include just anything. Mihandoost, picking up on Farhadi's affection for videocassettes, asks if he saw *Soltān-e qalb-hā* (King of hearts, dir. Mohammad Ali Fardin, 1968) on Betamax as a young child. *King of Hearts* was a popular romance film, directed by and starring heartthrob Mohammad Ali Fardin. Farhadi quickly retorted, "No. I saw part of it later. Apparently, whoever gave my father the [video] player was a person with good taste [*bāsaliqeh*]. He also gave him good movies." In Farhadi's account, video-cassettes preserved a certain kind of movie culture—the kind that would spur on a young filmmaker like himself.

Mihandoost's interview consciously connects the circulation of vid-eocassettes to Farhardi's generation of filmmakers. In his introduction, Mihandoost explains his motivations for the book, claiming that Farhadi is an exceptional example from "the generation of filmmakers who entered filmmaking after the revolution."[30] Throughout the first section of the interview, Mihandoost ties the history of videocassettes in Iran to Farhadi's own biography. When Farhadi references his university days in the mid-1990s, for example, Mihandoost asks, "At that time, videocassettes had become VHS?" For Mihandoost, Farhadi represents a curious "generation who got to know cinema by watching classic and contemporary films on video."[31] He asks Farhadi how he developed nostalgia for old Iranian cin-ema when his age prevented him from watching those films in the movie theater. He replies, "I watched some of them as an adolescent. As I became familiar with dramatic literature in my university years, I could distinguish between good and bad films. But I liked the music from *King of Hearts*, so I used it in a scene in *Shahr-e zibā* [Beautiful city, 2004]."[32] Farhadi's answer connects the dots of Mihandoost's theory. Old Iranian films were archived and accessed on videotape, and the movie culture that videos preserved would shape the next generation of filmmakers, too. The underground video infrastructure had preserved film culture by enabling access to even those films that the government had tried to eliminate.

The new generation of popular cinema contends with this relationship directly. In these films, video technology itself serves to mediate emotions, memories, and old modes of representation. Mani Haghighi's recent hit *50 Kilos of Sour Cherries* (2015), for example, shows how video preserved old film traditions while fostering new viewership. Director Mani Haghighi's body of work, both before and after *50 Kilos of Sour Cherries*, situate him comfortably in a contemplative and absurdist movement within Iranian cinema. Later in 2015, however, he tried his hand at an entirely different kind of filmmaking, stating in interviews that "with *50 Kilos of Sour Cherries* I wanted to experiment with commercial cinema."[33] *50 Kilos of Sour Cherries* was wildly popular in Iran and ultimately landed as the fourth-best performing film of 1395 (2015–2016) in box-office sales.[34] It would likely have moved even further up if officials had not prematurely pulled it from screens in the summer of 2016, after several high-ranking religious officials made complaints about the movie.[35] By the time it was released on DVD a year later, it had been censored heavily.

What made *50 Kilos of Sour Cherries* controversial also made reviewers liken it to the filmfārsi movies from four decades ago: it is a light-hearted romantic comedy.[36] After main characters Davud and Ayda enter into a sham marriage, they slowly begin to fall in love. Important to the present discussion is the fact that Ayda's mother, Farangis, once starred in a filmfārsi movie called *Nakhlestān-e havas* (Palm grove of desire), directed by Ayda's father. Of course, *Palm Grove of Desire* is not an actual film but rather a fictional detail intended to service the plot. When Davud meets Farangis for the first time, he recognizes her from the movie. He later brings a copy of it on VHS to their house and all of them watch it together (figure 5.4). The trailer for the movie, which we as viewers see, even features Googoosh's hit "Hamsafar" (Companion, 1975), a song from the soundtrack of a film by the same name, also starring Googoosh and her husband at the time, Behrouz Vossoughi. While we never actually hear Googoosh's voice, the first few bars of the introduction inevitably remind viewers of the film. Just for good measure, the trailer's voiceover even uses the word for companion,

Figure 5.4
In the popular movie *50 Kilos of Sour Cherries*, the character Davud brings a copy of
Palm Grove of Desire on video to show Farangis. He and his classmates used to pass the
videocassette around at school.
Source: *50 kilu ābālu* (dir. Mani Haghighi, 2016).

hamsafar, as a final hint: "hamsafar, tanhā naro!" (companion, don't go
alone!). Ayda's father, shocked to see the movie after so many years, asks
Davud where he found a copy. Davud explains, "In elementary school, we
used to pass it around. Everyone loved it!"

In a strange twist, Haghighi announces his own father, Namat Haghighi,
as the director of cinematography in the imaginary VHS trailer for *Palm
Grove of Desire*. Namat Haghighi was, in fact, a well-known cameraman,
who served as the director of cinematography for the film *Companion* and
a number of other popular movies from before the revolution, when Mani
Haghighi was still a young child. In this sense, Haghighi is continuing the
work his father did in popular cinema. There is further textual evidence to
support *50 Kilos of Sour Cherries* as the inheritor of the filmfārsi tradition.
It shares a number of overtones with *Palm Grove of Desire*, including a main
character named Davud and two unrequited lovers.

The videocassette in *50 Kilos of Sour Cherries* mediates this relationship
between Mani Haghighi's new venture into commercial cinema and the
older tradition of popular cinema, which had been his father's purview.
While all of these references may not have been readily apparent to viewers,
especially those who just wanted a good laugh, they nevertheless construct
video as a uniquely Iranian phenomenon. From trading tapes at school to

filmfārsi to Iran's biggest diva, Googoosh, the VHS cassette in *50 Kilos of Sour Cherries* embodies a prehistory that only makes sense in Iran. While earlier discourses on video in Iran position it as a technology of cultural imperialism from the West, as I've shown here, it was actually a protector of Iran's cultural heritage. Like *Sperm Whale 2*, *50 Kilos of Sour Cherries* transfers the private experience of watching movies on videocassette to the big screen. But unlike the *Sperm Whale* movies, which only deal with videocassettes in flashback, it understands video technology not only as a thing of the past but also as a technology of now, even if it has fallen out of mainstream use. The videocassette not only preserved *Palm Grove of Desire*, an imaginary piece of filmfārsi history, but also found new viewership for it in Davud, his classmates, and the real audiences who saw *50 Kilos of Sour Cherries*.

In the same spirit as *Sperm Whale* and *50 Kilos of Sour Cherries*, Mostafa Kiayee's *Bulletproof* shows continuity between the past and present of movie culture. Rather than underscore filmfārsi content like *50 Kilos of Sour Cherries*, *Bulletproof* foregrounds the informal labor and processes that have determined how movies circulate. In particular, it traces a line of continuity between videocassette distribution and more contemporary informal distribution practices, especially the circulation of movies on VCD. Like the other films discussed in this section, *Bulletproof* belongs to the new popular cinema. It was one of the top-selling films of 1391 (2012–2013),[37] and reviewers celebrated it at the time as innovative and as a rare instance in which an Iranian satire actually made audiences laugh.[38] The director has stressed his desire to entertain audiences more than anything. At the 2016 Fajr Film Festival in Tehran, he claimed, "I am neither a philosopher of film nor a fan of slogans. My cinema is a cinema of entertainment." He continued, "I prefer to leave the deep sentiments to the other directors in Iranian cinema. I want people to enjoy my movies."[39] Kiayee has found success in this particular kind of cinema and has released a successive line of popular movies since the early 2010s.

Bulletproof, his second feature film and first breakout hit, grapples with the long history of video in Iran. It begins in the 1980s at the height of the Iran-Iraq War. It focuses on Salim, a middle-aged man who makes a modest living by duplicating and distributing movies on video and Persian-language pop music from Los Angeles on audiocassette. When Salim discovers that he has cancer and only two months to live, he decides to atone for his sins—specifically his work with illegal cassettes—by enlisting in the war, where he hopes to die a martyr. Unexpectedly, he survives both cancer and the warfront. The movie ends several years later with Salim in his old office: a room filled with video equipment. This time, however, he pulls freshly copied VCDs out of the player rather than a videocassette. Salim lives another day and continues his work in underground video distribution. This is a departure from earlier representations of video dealers like *Song of Tehran* (1992), in which informal video distribution is condemned through the death of the video dealer character.

Bulletproof not only spares Salim an untimely death but also saves video from obsolescence. In the final scene, Salim's business, converted from analog to digital, shows how the networks that were forged through videocassettes later supported digital media like VCDs and DVDs. The film concludes with Salim grabbing his bag full of discs and hitting the street to return to work. The technology may have changed, as emphasized by careful close-up shots of Salim's computer tower, but the routes and labor have not. *Bulletproof*, through its representation of video, thus reminds viewers that (1) the system of underground circulation is nothing new and, perhaps more important, that (2) even as media seem to become more ephemeral and less material, they still depend on labor and infrastructure to exist.

Taken together, examples like *Sperm Whale, 50 Kilos of Sour Cherries*, and *Bulletproof* showcase the refusal of media to go away. The underground video infrastructure of the 1980s and 1990s is curiously present, as Iranian society redefines what popular cinema can or should do in the twenty-first century. That videocassettes continue to occupy a place in Iranian media provides an important corrective to the way in which media history has been conceptualized—as stories of innovation and revolution, of abrupt starts and finishes, and of the life and death of technologies. Media and the social practices that develop around them do not just die, even if the technology that supports them falls out of favor or is no longer considered mainstream. The way in which recent Iranian films contend with the history of videocassettes invites us to clear out a space to think about the afterlives of media technologies, not as progress or "identity crisis,"[40] but as structures, practices, and objects that are supposed to be dead but still exist in some form.

AN ARCHIVE FOR THE AGES

The underground video infrastructure does not just persist in the fictional worlds of works like *Sperm Whale* and *50 Kilos of Sour Cherries*. It also continues to exist in material and cultural practices, as real people manage, collect, and archive the videocassettes that once animated underground distribution channels. As the narrative about the apartment video stash at the beginning of this chapter demonstrates, video still matters in Iran in the most material way. Videocassettes as material objects continue to structure how people remember and relate to underground video culture. Popular cinema and social media posts may render the history of video visible and transform it into immaterial sights and sounds. But those attempts to memorialize video culture lie grounded in the heaps, piles, and bundles of cassettes that still exist in the country.

As they remember video, Iranians often refer to their orderly stacks of videos as an archive (*ārshiv*). Of course, archives constitute the site at which remembering becomes material. And yet this seems a curious designation, since my interlocutors were not referring to the formal archives of public institutions but rather to their own personal collections or those of others. What does it mean to archive? And who has permission to do it? If the video infrastructure, by and large, existed outside of formal institutions, then where are its archives? Looking at how people talk about, write about, and represent video "archives," I challenge the conventional wisdom on media collecting, which pivots on capitalist-driven consumer culture, especially in the United States. Instead, the personal video collections that exist in Iranians' basements and in their imaginations embody an effort to participate in the collective memory of an underground world that was never supposed to exist.

To talk of a video archive in Iran is to trip over a paradox. An archive of videocassettes is contradictory in three ways. First, consumer video was never meant for permanent storage. As I discussed in chapter 2, the technology's hardware breaks down over time and with repeated use. Even an untouched, well-stored videocassette suffers from magnetic remanence decay to its ribbon after several years. While preservationists debate whether celluloid film or digital formats can best preserve cinema, none would argue that videocassette technology is the ideal long-term home for media content. This fact challenges a branch of scholarship on media collecting that argues collectors are seeking a higher level of quality in purchasing media that have fallen out of mainstream use, such as vinyl records.[41] Second, a video archive raises the question of what is being archived. One typically thinks of archives as material forms but also as the information they store. As I just noted, media content cannot survive indefinitely on analog video. Plus, very few of the Iranians with whom I spoke still own a video player, even if they have cassettes. Perhaps most important, though, given the rate of magnetic remanence decay, the very glitches—the blips and blurs—that had made the experience of watching a movie on video so pleasurable would no longer exist in the same way. What is being collected, then, is the material object itself. But what is the benefit of collecting hundreds or thousands of objects that are identical in their hardware? Third, and finally, the video archives that exist in Iran aren't really archives at all—at least not in the strictest definition of the word, which foregrounds formal and institutional efforts to preserve information. Yet, as I will show, people use this designation to intervene in a history that was never entirely public or institutional in the first place.

Attempts by the state to archive video technology are meager at best. Such efforts certainly do not speak to the underground video culture that the government never sanctioned. The National Film Archive of Iran (NFAI), which is housed in the Ministry of Culture and Islamic Guidance (MCIG) building in Tehran, include a large room devoted to storing video-cassettes. I visited this room on a tour of the NFAI facilities. Unlike other divisions in the building, which included painstakingly detailed descriptions of preservation processes, archive officials had very little to say about the material stored there. According to my guide, the room was a holdover from the days of video technology. Now the cassettes just sat collecting dust, he told me. The videocassettes were not organized in any discernible way and included both Iranian and foreign movies. Most of the collection appeared to be copies rather than studio-produced cassettes; they had hand-written labels and generic sleeves. Just as the state had venerated cinema on 35mm film during its first two decades, the NFAI now also prioritizes the preservation of movie culture as celluloid film.

Similarly, while the Cinema Museum in Tehran displays relics of film technology, including film cameras, reels, and canisters, it provides no evidence of the video technology that sustained movie culture for several decades. Instead, as one tribute to video put it, "At the Cinema Museum, you can go downstairs to a room dedicated to the shows we would watch on video as children." In this instance, the Cinema Museum prioritized what people were watching rather than how they were watching it. In other words, it has stripped the video culture of the materials and technologies that made it special. Ultimately, it makes sense that public institutions like the NFAI and Cinema Museum would neglect videocassettes in their archiving practices—especially given the informal, underground nature of the video infrastructure of the 1980s and 1990s.

In the absence of these formal archives devoted to video, individuals have used collecting to preserve the history of the underground. When I was in Tehran studying analog video—a fallen media technology that had existed underground for most of its history in Iran—I expected to encounter the ephemeral: stories drenched in nostalgia, fading memories, and forgotten moments. Instead, I discovered a Tehran bursting with videocassettes. Hidden under beds, shoved behind video store counters, stacked in the back of vintage shops, and piled up in storage units, these video tapes have returned to the underground, where they continue to haunt the city. When I asked people why they continue to hold onto these videocassettes, their answers wavered between logistical and sentimental. Everyone, however, agreed on one thing: they couldn't bear to get rid of their old videocassettes, at least not just yet.

I want to return to the fact that people refer to these collections—whether their own or someone else's—as archives. This word has been central to many attempts to remember the underground video infrastructure. For example, Media Farzin's retrospective essay, "The Video People," which I cited in chapter 2, refers to personal collections as archives many times. Of her own video dealer in the late 1990s, she writes, "The very idea of Mister J's archive, and those like his, represented the determined pursuit of an intellectual tradition of film appreciation."[42] Herein lies the motivations for calling a personal collection of this kind an archive. These private video collections were about preservation as much as access; the collections were sites where people came together and produced knowledge about movies. Farzin continues, "It [the archive] was resolutely global. It was a precarious not-quite or only-occasionally public sphere, an amorphous discursive zone that moved nomadically from place to place, taking shape when and wherever it found a temporary audience."[43] In calling it an archive, the private video collection becomes a site of public knowledge and remembering.

These acts of "archiving" are not just discursive, they're material, too. Representations of these collections position them as assemblages where their thingness intermingles with the knowledge they produce. In such accounts, the videocassette archive resists new technologies at the same time that it challenges our knowledge of movie culture in Iran. Babak Tabarraee's novel *Kālt* (Cult, 2016), for example, reimagines the history of Iranian cinema. Its protagonist, Keyvan Shah'alidoost, is a university lecturer who has recently returned from Canada, where he was completing his graduate studies. He has been charged with teaching the course "History of Cinema 3," which he decides to teach as a history of Iranian cinema, much to his students' surprise. Because universities in Iran do not teach courses on Iranian cinema, Keyvan's course is the first in this fictional world. Recently divorced and living alone with his father, Kevyan only ventures outside of the home to teach and to visit his only friend, Amin, whose apartment has been taken over by an extensive videocassette archive.

In *Cult* the video archive is a space where new histories are possible and old histories cease to exist. As Keyvan describes it, people visit Amin not only because of his gentle disposition or his vast knowledge, but also

> because of the astonishing archive of films that Amin had collected since the arrival of video until the moment of its demise. Throughout the 1980s and 1990s, his old apartment in Darabad had gradually become blockaded with metal shelves, each housing tens or hundreds of VHS and Betamax tapes from the history of Iranian and world cinema. . . . Amin's very existence was tied to that archive, and the removal of even one of the thousands of cassettes wouldn't have been any different than someone strangling him to death.[44]

Amin's collection of videocassettes has become so entangled with his life that he refuses to leave it, instead insisting that people visit him inside his archive. During one such visit, Keyvan receives a mysterious call from a woman who tells him to report to an address to pick up the iPad that he left behind in his class. How the woman knew to call him there is as mysterious as her strange way of speaking. This call initiates an adventure that leads Keyvan to a remarkable discovery, as he learns about a sound film that predates *The Lor Girl*, generally believed to be the first Persian talkie film. Thus, what emerges from the video archive in *Cult* are new forms of knowledge production (like a university class on Iranian cinema), as well as new histories of cinema entirely.

My own motivations for writing this history of videocassettes in Iran began with one such archive. After my doctoral adviser, M. R. Ghanoonparvar, retired in 2013, he called one hot summer day and asked me to come to his house, just a few miles north of Austin, Texas. "I have something for you," he told me over the phone. When I arrived, he had carefully arranged a row of boxes for me in his garage. "It's an *ārshiv* [archive]," he explained. I packed the boxes in the car and rushed home, eager to see what was inside. What I found was an expansive collection of Iranian cinema on video. The boxes contained hundreds of tapes. Some had been distributed by companies in Los Angeles that specialized in circulating Iranian movies among the diaspora. Others had been taped off satellite or copied from some other source. As touched as I was by the gesture, I was also confused. At that point, I did not even have a TV, let alone a VCR. What was I going to do with all these cassettes? Ashamed to have them on full display in my office, I relegated them to a cabinet, where they could sit safely behind closed doors. They seemed too important to throw away, and I figured someday they might be a good resource if I ever needed to watch a film that I couldn't find somewhere else. The first time I tried to watch one of the videos, I rummaged through my department's conference room looking for a VCR. I found one hidden behind a stack of cables. The player immediately ate the cassette; I think it is still in there.

Though this archive did not immediately prove useful to me, I thought of it often. I had grown up with videocassettes in the 1980s and 1990s in the United States, and I wondered about the stories these tapes represented. What kind of networks had determined their circulation? Where had they come from, and why? What must it have been like to access and watch movies at home in Iran? What kind of communities were made and unmade through the exchange of plastic and ribbon? Little did I know that these questions would take me to an expansive underground world of video that I had never imagined could exist. The cassettes in my small, inherited archive led me to this world not because of what I saw on them but rather because they existed—bulky reminders of another time and place. I left my position at the University of Texas at Austin several years ago, and with it my collection of Iranian movies on video. I placed them in the guardianship of my students, hopeful that their presence would continue to spur new ideas and new ways of thinking about media in Iran.

CODA: WRITING VIDEO'S UNFINISHED HISTORY

While I was researching and writing this book, all roads seemed to lead me to Amir Ezzati. Although I did not know it at first, he was the anonymous author of the book *Vide'o va rasāneh-hā-ye tasviri* (Video and visual media), which had been an essential guide for me as I started to orient myself to the history of video distribution in Iran. I read the book so many times that the binding on my copy began to fall apart. Ezzati also came up often in my interviews. He was prominent for knowing just about everything there was to know about Iranian cinema. He had worked as a film critic, translator, journalist, distributor, and employee of the National Film Archive of Iran. His father had even owned a movie theater. Ezzati was perhaps most famous in the late 1990s and early 2000s for his expansive video archive. "If it existed on video, Ezzati had it," someone once told me. But that is all in the past now. Ezzati now lives in quiet Helsinki—far from his family, from the bustle of Tehran, and from the apartment that had once housed his videocassette collection.

It was not difficult to get in touch with Ezzati. Many of the people I spoke to were already in contact with him, and he graciously agreed to chat with me. More challenging was convincing him of the importance of my research. He began our first conversation by telling me that he thought it was strange I was studying the history of videocassettes, since people in Iran don't even value the study of Iranian cinema, a medium usually regarded as more legitimate. I nervously stumbled through some answer about how the means of access are always serious, no matter what people watch. "I'm curious," I told him, "to know how people used video to construct a world that existed underneath the state." "*Konjkāv*," he replied, correcting my pronunciation of the Persian word for "curious." I am not sure if his opening remarks were meant to be a test, but if they were, I'd like to think I passed.

In the exchanges that followed, I learned a lot about video in Iran from Ezzati; much of that information already adorns the preceding chapters.

But I also learned about what it took for him to write a history of video in *Video and Visual Media*. It is worth reflecting on that experience here, as it speaks to the larger concerns of this book. Even decades after the video ban, the state's desire for control still shapes media, including media scholarship.

Ezzati had just been released from prison when he got the request to write *Video and Visual Media*. The editors of a series called *Twenty-Five Years of Iranian Cinema* solicited him to write the volume. The request both surprised and frustrated him. As he would later write, "In a country where it was forbidden and considered a crime to have movies on video, even for me, a writer, film critic, television producer, and instructor of cinema, what was the point of publishing a book on such a topic, and why did I have to be its author?"[1] Ezzati's concerns echoed some of the tensions implicit in a mandate to write such a book in Iran. As I have documented throughout *Underground*, the state labored systematically to erase videocassettes from the public sphere. A book that sought to publicize video's tenacious presence came up against the structures, policies, and practices that had kept videocassettes invisible for so long. Plus, the people most qualified to write such a book were also the most vulnerable. They were, after all, the ones who had been active in the underground circulation of cassettes in the first place. It seemed that to write such a book, even ten years after the ban had been lifted, meant to step into a carefully set trap.

This predicament was especially true for Ezzati, who had been arrested just months earlier along with several other film critics. He was held in solitary confinement for sixty days before being released on steep bail. Among his charges was distributing illegal movies on video—an accusation he denies to this day. The circumstances of Ezzati's arrest and those of other journalists in 2003 remain shrouded in mystery. During this period, dozens of journalists and intellectuals were imprisoned, beaten, tortured, and forced into making confessions. This was a tense time in Iran. President Mohammad Khatami's promise of a free press and moderate cultural policies had backfired, and conservative members of the government held their ground by attacking the country's cultural institutions.

These arrests did not just include the cultural elite. One of the most well-known reactions to the string of arrests in 2003 was a letter entitled "A Letter from Tehran" by F. Parsa, which was translated and circulated widely online. In the letter, Parsa writes of the arrest of his friend, Mahmoud Vakili, who was not a well-known journalist like the others but was a cultural laborer nonetheless. As he explains, "Mahmoud Vakili fits nowhere in the 'system.' There was no crack or fissure through which he could gain entry. Finally, he became a 'filmi.' A term and an occupation which must not exist anywhere in this world other than in this wasteland. He collected films on tape and on DVD, threw them in his shoulder bag and rented them to people."[2] The arrests of Ezzati and Vakili reveal how dangerous video continued to be, even ten years after the formal ban was lifted.

Ezzati eventually agreed to write *Video and Visual Media*. He decided that no other possible candidates existed and, "if I didn't do it, then this historical project would be incomplete, which was at odds with the rigor and importance that all of us placed on accurately recording the history of our country's cinema."[3] He managed to write the book in three months. To research it, he scoured newspapers, magazines, and journals for references to home video technology and depended on unofficial sources like booklets published by video clubs before the ban. As he told me in an interview, however, the most important sources were his own experiences and notes, as well as those of the videocassette producers, distributors, and collectors he personally knew.

In the end, the editors of the *Twenty-Five Years of Iranian Cinema* series decided to play it safe. They published the book without Ezzati's name on the cover or in the other publication information. In the meantime, Ezzati was formally convicted of threatening national security as an agent of cultural imperialism. He fled to Turkey and eventually, through the UN High Commission on Refugees' resettlement program, landed in Finland, where he continues to reside today. He received a copy of his book before his abrupt departure, but he lost it in Turkey during one of his many moves. It wasn't until a friend brought him another copy eleven years after its original publication that he was finally reunited with his book. He scanned the copy and added a new introduction called "Hekāyat-e enteshār-e ketāb bedun-e nām-e mo'allef" (The story of publishing a book without the name of its author"). Then he republished it online in June 2016 through *Bāshgāh-e adabiyyāt* (Literature club), a Facebook page that Ezzati operates, which, as of December 2020, boasts nearly forty-five thousand members. In the introduction to the electronic version of *Video and Visual Media*, Ezzati writes: "Now that I am looking at this book, I find myself with a stranger. It is like I have found my long-lost child, and reconnecting is essential. We don't know each other very well, and we need other people to give their stamp of approval to believe it. That is why I am publishing an electronic version of this book: to triumph over that twelve-year nightmare."[4]

Ezzati's story about writing and publishing *Video and Visual Media* reveals the state's systematic attempt to undermine a comprehensive and cohesive history of home video technology in Iran. The book was published in 2004, a decade after the MCIG removed the video ban. Yet it was still a contentious topic. That Ezzati's name never appeared on his book and his ultimate exile from Iran represent other kinds of erasures, too. The story behind the publication of *Video and Visual Media* suggests how critics and scholars found themselves caught between wanting to acknowledge a history of video technology and the need to obscure certain parts of it, specifically the informal, illegal networks that sustained it for so long and the lasting impact it had on viewership in the country. This is strikingly visible in the text of the book itself. *Video and Visual Media* comprises four chapters: "Video: 1979–1983," "Video: 1983–1994," "Video and Visual Media," and "Institutions for the Production, Duplication, and Distribution of Video Programs." Even though the ban on video in 1983 organizes the structure of the book, the decade-long ban period (1983–1994), which constitutes the longest timeframe in the twenty-five years covered, receives only two and a half pages of commentary. This disparity in and of itself implies the challenges of writing about the underground video infrastructure.

In his new introduction to the book, Ezzati writes that during his imprisonment, despite his many objections to the contrary, his captors insisted, "You dumb ass, cinema is politics," as they beat him.[5] By extension, so too was the history of motion pictures political at the time. *Video and Visual Media* was published at a very distinct particular moment in Iran's political and cultural history, at the tail end of Mohammad Khatami's second term as president. As I have written elsewhere, Khatami's presidency was a period marked by attempts to engage more openly with the global community and to reorient state rhetoric to favor concepts like democracy and reform. Cinema played a big role in both these projects.[6] At the same time, Ezzati's and his colleagues' arrests—alongside smaller aggressions restricting how critics and scholars could represent the contemporary history of motion pictures—expose the paradoxes of Khatami's presidency. While Khatami claimed to support cultural initiatives and the press, complete intellectual freedom remained elusive. The question remains why. In light of Khatami's moderate cultural politics, why was the history and even the presence of analog video technology still problematic ten years after the lifting of the video ban and four years after it ceased to be a crime to possess an unpermitted film on video?[7]

The answer rests in media's refusal to go away. If we take a broad view of media that considers not just technologies but also the ways in which those technologies structure communities, communication, and practices, then video in 2004 was hardly history at all. The laws and policies that developed to regulate the distribution of media formed because of analog video technology. So, too, did the networks that allowed the rapid, albeit underground, circulation of motion pictures. As I showed in chapter 5, even the physical videocassettes themselves continued to exist. While digital technologies may have overcome their analog predecessors, the underlying systems of control, circulation, and even the materiality of video remained intact. As Ezzati constantly reminded me during our conversations, the story of video wasn't finished at the time he was writing it. The use of digital media and the internet were growing rapidly and uncontrollably in Iran,[8] and this reawakened old anxieties about the uncontrollability of media. I very purposely use the word "reawaken" here to emphasize that digital technologies did not activate new apprehensions for the state but rather summoned the prior concerns that formed around earlier iterations of unruly technologies like videocassettes. A book like *Video and Visual Media* constructed a narrative in which the illegality of media can be successfully contained and controlled. Such an approach played precisely into this moment in the early 2000s, which was fraught with the state's anxieties about technology, media, and culture.

A MATTER OF MEMORY

In the spring of 2016, shortly after returning home from Tehran, I received an email from a friend there. "It seems," he reported, "that a London-based filmmaker is here now, interviewing some of the same people as you and asking similar questions about videocassettes." The enthusiasm that my research trip had inspired suddenly turned sour. Typical of a researcher recently returned from the field, I thought I had stumbled across something entirely unique.

I thought it was my history for the taking. With trepidation I reached out to Niyaz Saghari to learn more about her film and to see what it would mean to try to collaborate. What I got in reply was a renewal of my enthusiasm and a touch of humility—affirmation that this was an important story to tell and that it could be told in a variety of ways. Saghari shared a rough cut of her film as soon as it was available. Partly an attempt to record this history before it disappears and partly a nostalgic engagement with the past, *VHS Diaries* embodies all the modes of remembering discussed in chapter 5.

VHS Diaries is an experimental documentary divided into two parts. In the first, statements from anonymous experts establish the unique culture of video distribution in the 1980s and 1990s. In the second, Saghari focuses her story on one video dealer, whom she calls Mr. J. In this part, we learn about Mr. J, his motivations for dealing in videocassettes, and his time in prison because of it. He even reenacts his old line of work: creating labels for new cassettes, packing them in a large briefcase, donning his old trench coat, and then slipping into a car where a driver surreptitiously carries him through the streets of Tehran. Saghari would later tell me that Mr. J had been her filmi and inspired her to make such a documentary. Indeed, *VHS Diaries* is very much Saghari's own diary of participation in underground video. Images of the filmmaker, her credentials (like membership cards), and even pages from her own journals remind us that this is her story as much as the story of Mr. J and his videocassettes. Haunting scenes of VHS tape blowing through stairwells and empty city streets and shots of abandoned buildings far outside the city limits pepper the documentary. They begin the work of theorizing what it means to remember a history of video in Iran. Although the presence of tape on screen reminds us of the videocassette's materiality, it has now been reduced to a shape less than a millimeter thick. It is a shape that threatens to blow away, the camera barely able to keep up. Meanwhile, the scenes outside of the city, with its tight streets and interior spaces, deliver us to dead ends and dark places. When taken together, these disruptive scenes invite speculation on where such attempts to remember video take us and what impact they have on its history.

"Memory is a kind of photographic film," Mary Carruthers writes, "exposed . . . by an amateur and developed by a duffer, and so marred by scratches and inaccurate light-values."[9] Such a claim shows how even a medievalist like Carruthers might find value in an anachronistic metaphor that likens memory to some modern media technology. This book, however, has turned that metaphor on its head and asked instead what it means to remember a medium itself. The answer to such a question, of course, includes the memories or pieces of history that the medium helped facilitate or preserve. Yet it also demands much more. It considers what other media have been implicated and how these acts of remembering a medium unsettle those easy histories of media that have tidy beginnings and ends and do not account for the "scratches and inaccurate light-values" of memory. In the previous chapter, I claimed that various acts of remembering—including collecting, recording, and representing—together insist on a unique history of video in Iran. But what is the history that has been shaped and colored through these memory practices?

The history of video that remembering has created in Iran exists in the sways of the senses and the emotions, between what can be observed and what can be felt. That said, it is above all else a material history. Despite the convenient binary that I have just set up and the easy turn of phrase, I do not necessarily wish to evoke Marx's "historical materialism" by calling this a material history, although one could certainly draw important parallels. Rather, I want to emphasize video technology's matter, which has not just shaped but has also determined its very historical consciousness in Iran. Video technology, through its physical artifacts like cassettes and players, marked a materialization of moving images for a large segment of society, and as such it became a site where senses and emotions intersected in unprecedented ways. People could touch and hold the same movies that stirred their emotions. The fact that some Iranians currently refer their dusty boxes of videocassettes as archives helps us appreciate further how these objects serve as touchpoints for the history of video. As soon as the physical artifacts disappear, so too might the memories associated with them, which is why people continue to hold on to them long after their value as usable objects has diminished.

Even my writing this book has been an act of remembering, as I have intervened in remembering video in Iran. By conducting oral history interviews, I have intentionally activated memories and insisted that the history of video remain present for those Iranians with whom I interacted. However, the anonymity of my research—the fact that I followed procedures to protect the identities of my interlocutors—has meant that my own encounters with this history are now also trapped as fragmented memories. They reside in disembodied voices on audio file, photographs that people sent to me, and my own attempts to fill in the blanks. I, too, find myself caught between wanting to record the experience of this history as accurately as possible and acknowledging that remembering is always somehow incomplete. The fact that video continues to be discussed and contested—the fact that it continues to be remembered—tells us what an important medium it was and continues to be.

NOTES

INTRODUCTION

1. Joshua M. Greenberg, *From Betamax to Blockbuster: Video Stores and the Invention of Movies on Video* (Cambridge, MA: MIT Press, 2008), 17–40.

2. Caetlin Benson-Allott, *Killer Tapes and Shattered Screens: Video Spectatorship from VHS to File Sharing* (Berkeley: University of California Press, 2013), 203.

3. See, for example, Greenberg, *From Betamax to Blockbuster*; Daniel Herbert, *Videoland: Movie Culture at the American Video Store* (Berkeley: University of California Press, 2014).

4. Peter Decherney, *Hollywood's Copyright Wars: From Edison to the Internet* (New York: Columbia University Press, 2012), 155–200; Frederick Wasser, *Veni, Vidi, Video: The Hollywood Empire and the VCR* (Austin: University of Texas Press, 2002).

5. Ramon Lobato, *Shadow Economies of Cinema: Mapping Informal Film Distribution* (London: BFI/Palgrave Macmillan, 2012); Ramon Lobato and Julian Thomas, *The Informal Media Economy* (Cambridge: Polity Press, 2015).

6. My thinking here is indebted to Brian Larkin's work on media infrastructures in Nigeria. In particular, he argues that we should take seriously the infrastructures that form out of technical and regulatory failures. Brian Larkin, *Signal and Noise: Infrastructure and Urban Culture in Nigeria* (Durham, NC: Duke University Press, 2008).

7. For more on media during the revolution, especially consumer technologies, see Annabelle Sreberny-Mohammadi and Ali Mohammadi, *Small Media, Big Revolution: Communication, Culture, and the Iranian Revolution* (Minneapolis: University of Minnesota Press, 1994).

8. For more on how the country's war veterans shaped state-controlled media production, see Narges Bajoghli, *Iran Reframed: Anxieties of Power in the Islamic Republic* (Stanford, CA: Stanford University Press, 2019).

9. As the term "home video" suggests, the home was central to the underground and the negotiations that took place there. The video-capable home operated as a place where private and public interests intersected and contradicted, repelled and melded. This is consistent with what we know about home viewership elsewhere.

See, for example, Barbara Klinger, *Beyond the Multiplex: Cinema, New Technologies, and the Home* (Berkeley: University of California Press, 2006).

10. One analogous example is the case of Cuba, where officials forbade the sale of VCRs until 2008. Even though this measure made accessing home video difficult, owning or using a VCR was never illegal, a fact that distinguishes the Cuban case from its Iranian counterpart. For more, see Anna Cristina Pertierra, "Private Pleasures: Watching Video in Post-Soviet Cuba," *International Journal of Cultural Studies* 12, no. 2 (2012): 113–130.

11. Fyodor Dostoyevsky, *Notes from the Underground*, trans. Richard Pevear and Larissa Volokhonsky (New York: Vintage, 1994).

12. Rosalind Williams, *Notes on the Underground: An Essay on Technology, Society, and the Imagination* (Cambridge, MA: MIT Press, 2008).

13. Karen Barad, *Meeting the Universe Halfway: Quantum Physics and the Entanglement of Matter and Meaning* (Durham, NC: Duke University Press, 2007): 139.

14. Michel Foucault, *An Archeology of Knowledge*, trans. A. Sheridan Smith (New York: Pantheon Books, 1972), 117. My claim here builds off Foucault's concept of discourse. He refers to a discursive practice as "a body of anonymous, historical rules, always determined in the time and space that have defined a given period, and for a given social, economic, geographical, or linguistic area, the conditions of operation of the enunciative function." Foucault's work provides a useful starting point for the theoretical ambitions of my study of the underground. However, I ultimately defer to Barad's definition of material-discursive phenomena. I share in her concerns about Foucault's position on materiality. She argues that Foucault has failed to "articulate . . . the precise nature of the relationship between discursive practices and material phenomena" (Barad, *Meeting the Universe Halfway*, 200). While Foucault insists on a distinction between human and nonhuman bodies, Barad argues in favor of intra-action, wherein agency is a dynamic force shared among different material bodies, human and nonhuman alike. In her account, "human practices are not the only practices that come to matter" (223). Thus, a material-discursive apparatus helps illuminate the ways in which the materiality of home video technology helped constitute the underground as a reality in Iran in the 1980s and 1990s. As Barad writes, "Reality is an ongoing dynamic of intra-activity" (223). For more, see Barad, *Meeting the Universe Halfway*. In *Underground*, I map such moments of intra-activity and chart the various materialities that constituted the underground at any given moment.

15. Lucas Hilderbrand, *Inherent Vice: Bootleg Histories of Videotape and Copyright* (Durham, NC: Duke University Press, 2009), 15.

16. We might, for example, consider Pamela Karimi's examination of team houses (*khāneh-hā-ye timi*) under Mohammad Reza Shah (r. 1941–1979). These were anti-Shah—often communist—cells that used normal family dwellings to organize

their political efforts. Of these homes, Karimi writes, "A simple family unit, for instance, could accommodate a 'threshold' in the kitchen cupboard from which one could leave the 'outer' house and enter an 'inner' home, where the most sensitive materials and documents were kept, and the most important meetings were held." Karimi importantly shows how the same logic that transformed the domestic sphere into underground spaces has continued to inform an "ethos of design," wherein "concealment" is a central tenet. For more, see Pamela Karimi, "Alternative Belongings: Instituting and Inhabiting the Iranian Underground" in *After Belonging: The Objects, Spaces, and Territories of the Ways We Stay in Transit*, ed. Lluís Alexandre Casanovas Blanco et al. (Zurich: Lars Müller Publishers, 2016), 98–109.

17. Mahmood Shahabi, "The Iranian Moral Panic over Video: A Brief History and Policy Analysis," in *Media, Culture and Society in Iran*, ed. Mehdi Semati (London: Routledge, 2007), 111–129. This chapter is one of the few scholarly works that deals directly with this topic.

18. Lisa Parks and Nicole Starosielski, "Introduction," in *Signal Traffic: Critical Studies of Media Infrastructures*, ed. Lisa Parks and Nicole Starosielski (Urbana: University of Illinois Press, 2015), 5; Alissa Perren, "Rethinking Distribution for the Future of Media Industry Studies," *Cinema Journal* 52, no. 3 (Spring 2013): 165–171.

19. Catelin Benson-Allott, *The Stuff of Spectatorship: Material Cultures of Film and Television* (Berkeley: University of California Press, 2021); Benson-Allott, *Killer Tapes and Shattered Screens*; Anna McCarthy, *Ambient Television: Visual Culture and Public Space* (Durham, NC: Duke University Press, 2001); Lisa Parks, "'Stuff You Can Kick': Toward a Theory of Media Infrastructures," in *Between Humanities and the Digital*, ed. Patrik Svensson and David Theo Goldberg (Cambridge, MA: MIT Press, 2015), 355–373. My work is also indebted to a tradition of scholarship on media reception, especially Ann Gray, *Video Playtime: The Gendering of a Leisure Technology* (London: Routledge, 1992).

20. Decherney, *Hollywood's Copyright Wars*, 1.

21. Here we might see resonance with Robert Ellickson's seminal work *Order without Law: How Neighbors Settle Disputes* (Cambridge, MA: Harvard University Press, 1991), in which he argues that laws are largely incidental, and that people develop social norms and practices to govern themselves.

22. Bajoghli, *Iran Reframed*; Kevan Harris, *A Social Revolution: Politics and the Welfare State in Iran* (Berkeley: University of California Press, 2017); Manata Hashemi, *Coming of Age in Iran: Poverty and the Struggle for Dignity* (New York: New York University Press, 2020); Nazanin Shahrokni, *Women in Place: The Politics of Gender Segregation in Iran* (Berkeley: University of California Press, 2019).

23. See, for example, Adrian Athique, "The Global Dynamics of Indian Media Piracy: Export Markets, Playback Media, and the Informal Economy," *Media,*

Culture & Society 30, no. 5 (2008): 699–717; Jonathan Gray, "Mobility through Piracy, or How Steven Segal Got to Malawi," *Popular Communication: International Journal of Media and Culture* 9, no. 2 (2011): 99–113; Hilderbrand, *Inherent Vice*; Larkin, *Signal and Noise*; Tristen Mattelart, "The Changing Geographies of Pirate Transnational Audiovisual Flows," *International Journal of Communication* 10 (2016): 3503–3521; Laikwan Pang, *Creativity and Its Discontents: China's Creative Industries and Intellectual Property Rights Offenses* (Durham, NC: Duke University Press, 2012); Shujen Wang, *Framing Piracy: Globalization and Film Distribution in Greater China* (Lanham, MD: Rowman & Littlefield, 2003).

24. I am reminded of Pang's caution that we must be careful not to assess bootlegging practices through the rubric of liberal, leftist politics. Pirated material may appear to subvert regimes of intellectual property, but it is also almost always driven by a desire for the commodity itself. In other words, even when piracy efforts are not in direct opposition to intellectual property regimes, they are still understood through the lens of commodity culture. For more, see Pang, *Creativity and Its Discontents*, 198.

25. Siva Vaidhyanathan, *Copyrights and Copywrongs: The Rise of Intellectual Property and How It Threatens Creativity* (New York: New York University Press, 2003).

26. Barbara Klinger, "Contraband Cinema: Piracy, *Titanic*, and Central Asia," *Cinema Journal* 49, no. 2 (Winter 2010): 106–124.

27. Ramon Lobato, "The Paradoxes of Piracy," in *Postcolonial Piracy: Media Distribution and Cultural Production in the Global South*, ed. Lars Eckstein and Anja Schwarz (London: Bloomsbury Academic, 2014), 124.

28. For more on the interplay between intellectual property right law and the circulation of Iranian cinema, see Mahsa Salamati, "Transnational Film Circulation in the Iranian Context: From Conjunctural Crisis to Discursive Heterotopia," Ph.D. diss., University of New South Wales, 2019.

29. Michael Z. Newman, *Video Revolutions: On the History of a Medium* (New York: Columbia University Press, 2014), 2.

30. Larkin, *Signal and Noise*, 3.

31. My thinking here has been influenced by Rita Felski, especially her claim that critique is both a mood and a method. For more, see Felski, *The Limits of Critique* (Chicago: University of Chicago Press, 2015).

32. Svetlana Boym, *The Future of Nostalgia* (New York: Basic Books, 2001).

CHAPTER 1

1. See, for example, Hamid Naficy, *A Social History of Iranian Cinema, Volume 3: The Islamicate Period, 1978–1984* (Durham, NC: Duke University Press, 2012); Hamid

Naficy, *A Social History of Iranian Cinema, Volume 4: The Globalizing Era, 1984–2010* (Durham, NC: Duke University Press, 2012); and Bajoghli, *Iran Reframed*.

2. Amir Ezzati, *Vide'o va resāneh-hā-ye tasviri* [Video and visual media], Bist-o-panj sāl-e sinemā-ye Iran [Twenty-five years of Iranian cinema] (Tehran: Cinema Museum, 2003), 19.

3. Ezzati, *Video and Visual Media*, 13.

4. Greenberg, *From Betamax to Blockbuster*, 63–80; Herbert, *Videoland*, 25–27.

5. Ezzati, *Video and Visual Media*, 18.

6. Ezzati, 18. The presence of Iranian films on these lists provides an important corrective to accounts that suggest there was a complete rupture in Iranian cinema with the revolution. There was formal, legal continuity through the circulation of videocassettes. Most of these films did not have permits to be screened in theaters but did circulate legally on video. This tells us that at the time the state was distinguishing between home viewing and public screenings, with more leeway in what could be distributed for consumption in the home. Video distributors were required to put an image of the designated permit at the beginning of the movie, thus creating a way to distinguish between permitted and unpermitted movies.

7. Ezzati, *Video and Visual Media*, 15.

8. See, for example, Peter Chelkowski and Hamid Dabashi, *Staging a Revolution: The Art of Persuasion in the Islamic Republic of Iran* (New York: New York University Press, 1999).

9. Newman, *Video Revolutions*, 36–46.

10. See, for example, Adrian Hope, "Sistem-e kāset-e khānegi-ye zabt-e barnāmeh-hā" [The home cassette system for recording programs], trans. Rahim Qasemian, *Tamāshā* (June 17, 1978): 16–17, 72.

11. Sreberny-Mohammadi and Mohammadi, *Small Media, Big Revolution*, xvii–xxiii.

12. Mark Thomas, "Homevideo Piracy, Penetration Killing Middle East Cinema," *Variety*, May 25, 1988, 76.

13. Evidence of these anxieties can be found in "Policies and Procedures for the Production, Distribution, and Exhibition of Films and Video Programs, and Supervision over the Import of Video Equipment," the first master-set of policies distributed by the Office for Video Affairs in the MCIG in 1995. In this document, the MCIG declares that one of the distinctive qualities of home video, in comparison to the television and film industries, was that it could be "produced and distributed quickly according to the demands of the audiences." For more, see "Policies and Procedures for the Production, Distribution, and Exhibition of Films and Video Programs, and Supervision over the Import of Video Equipment," in Ezzati, *Video and Visual Media*, 36.

14. For more on the concept of an Islamicate cinema, see Naficy, *A Social History of Iranian Cinema, Volume 3.*

15. Hossein Ghazian, "The Crisis in the Iranian Film Industry and the Role of the Government,". in *New Iranian Cinema*, ed. Richard Tapper (London: I. B. Tauris, 2002), 77–85.

16. Ghazian, "The Crisis in the Iranian Film Industry," 77.

17. See, for example, "Filmsāzi va emkānāt-e vide'o" [Filmmaking and the possibilities of video], *Sinemā dar vide'o* [Cinema on video] 1 (1982): 105; and "Tārikhcheh-ye vide'o va navār-e vide'o" [The history of video and the videocassette], *Sinemā dar vide'o* [Cinema on video] 1 (1982): 100–101.

18. "Khatar-e vide'o barāye sinamā cheqadr jeddi-st?" [How serious is video's threat to cinema?], *Sinemā dar vide'o* [Cinema on video], 2 (1982): 91.

19. Ezzati, *Video and Visual Media*, 19.

20. "Sokhanān-e vazir-e ershād-e eslāmi dar mored-e lāyeh-ye matbu'āt, tablightāt-e khārej az keshvar, taklif-e honarmandān, turism, tarh-e ettelā'āti-ye keshvar va taghirāt dar vezārat-e ershād" [Comments by the minister of culture and Islamic guidance about the press, advertising outside of the country, the work of artists, tourism, the role of information in the country, and changes in the ministry of Islamic guidance], *Ettelā'āt* (Tehran), December 12, 1982.

21. Ezzati, *Video and Visual Media*, 21.

22. Hamid Naficy, *A Social History of Iranian Cinema, Volume 1: The Artisanal Era, 1897–1941* (Durham, NC: Duke University Press, 2012), 59–60; Naficy, *A Social History of Iranian Cinema, Volume 2: The Industrializing Years, 1941–1978* (Durham, NC: Duke University Press), 264; Naficy, *A Social History of Iranian Cinema, Volume 4*, 345.

23. "Tasvir-e ebtezāl mahv mishavad" [The image of obscenity fades away], *Sorush* 194, June 4, 1983, 32–35.

24. I've searched widely for such documents and have not been able to find any. Of course, citing an absence is always inherently problematic to a certain extent, but the fact that religious rulings are not referenced indicate that these statements were driven more by political than religious concern.

25. Ervand Abrahamian, *A History of Modern Iran* (Cambridge: Cambridge University Press, 2008), 186.

26. Ali Khamenei, *Farhang va tahājom-e farhangi* [Culture and cultural assault] (Tehran: Sāzmān-e madārek-e farhangi-ye enqelāb-e eslāmi, 1994).

27. Evidence can be found in statements by Ali Larijani, the head of the MCIG in 1994. In an interview at the time of the legalization of video, entitled "Cultural Assault and the Will of a Nation," Larijani attempted to rescue video technology

from the negative connotations of cultural assault. He claimed that although video was a Western technology, it could be used as a means of resisting cultural imperialism rather than giving in to it. He suggested that in order to be competitive globally, both economically and ideologically, Iran needed not to reject technological advances but rather to adapt them in a way that suited the ideals of the Islamic Republic. Elsewhere the policies created by the MCIG after legalization identify the need for video institutions to "create aggressive programs against cultural assaults and corrupt foundations of foreign cultures and their distortions of Islam and the Islamic Republic." The notion of a cultural assault structured much of the commentary on video at the time of its legalization. This attention shows how important anti-imperialist rhetoric was to the official discourse on video during the ban. For more, see "Tahājom-e Farhangi va 'azm-e melli" [Cultural assault and the will of a nation], *Film* 135 (January 1992): 6; and Ezzati, *Video and Visual Media*, 38.

28. Ezzati, 13.

29. "Vide'o va ma'muriyat-e ān dar Iran" [Video and its mission in Iran], *Sorush* 338, May 31, 1986, 42.

30. "Qānun-e nahveh-ye residegi beh takhallofāt va mojāzāt-e forushandegān-e lebās-hā'i keh estefādeh az ānhā dar malā'-e 'ām khalāf-e shar' ast va yā 'effat-e 'omumi rā jarihehdār mikonad" [Law dealing with the violations and punishments of vendors of clothing whose use in public spaces is against religious law or violates public chastity], last modified March 18, 1987, Islamic Parliament Research Center of the Islamic Republic of Iran, http://rc.majlis.ir/fa/law/show/91340.

31. "Āmār-e vaz'iyat-e sinemā, televiziyun va vide'o dar 42 keshvar-e jahān" [Statistics on the status of cinema, television, and video in 42 countries around the world], *Film* 135 (January 1993): 8.

32. "Vide'o: Az zendegi-ye makhfi tā hozur-e āshkār" [Video: From a hidden life to an open present], *Vide'o-māhvāreh* [Video-Satellite] 1 (March 1994): 20–21.

33. "Iranian Film Biz Revisited: Lotsa U.S. Cassettes, Picture Backlog," *Variety*, June 6, 1984, 91.

34. Tahmasb Solhju, "Khub, bad, mofid, mozer, zesht, zibā, sāzandeh, virāngar" [Good, bad, useful, harmful, ugly, beautiful, constructive, destructive], *Film* 157 (April 1994): 97–99.

35. "Mas'aleh-ye vide'o va moshkelāt va 'avārez-e farhangi-ye ān" [The issue of video and its cultural problems and duties], last modified June 30, 1992, Islamic Parliament Research Center of the Islamic Republic of Iran, http://rc.majlis.ir/fa/law/show/100181.

36. Ezzati, *Video and Visual Media*, 37.

37. "Iranian Film Biz Revisited," 91.

38. Deborah Young, "Iran Unveils New Film Threat," *Variety*, February 17, 1992, 93.

39. Mark Thomas, "Iranians Just Want to Have Fun," *Variety*, February 14, 1990, 7.

40. Thomas, "Iranians Just Want to Have Fun," 7. Even as publications like *Variety* provide us with valuable information, we should be skeptical of them. After all, *Variety* is an American periodical that represents the interests of Hollywood. With titles like "Iranians Just Want to Have Fun," its articles played into a larger narrative at the time about an Iran whose oppressive policies stifled its citizens. And yet some of the articles' details provide a much more nuanced portrait of how and why people negotiated the legal framework of media regulation at the time. What emerges from these reports are the kinds of paradoxes and contradictions that formed the underground itself.

41. Samaneh Farahani, "Negātiv-hā'i keh sukhtand: negāhi beh ejrā-ye tarh-e jam'āvari-ye film-hā-ye khāreji" [The negatives that were burned: A look at the implementation of the plan to collect foreign films], *Shahr-e Ketāb* 6 (January 2015): 70.

42. Farahani, "The Negatives That Were Burned," 70.

43. Naficy, *A Social History of Iranian Cinema, Volume 3*, 82.

44. Farahani, "The Negatives That Were Burned," 71.

45. For more, see Naficy, *A Social History of Iranian Cinema, Volume 3*, 3–4.

46. Farahani, "The Negatives That Were Burned," 72.

47. Naficy, *A Social History of Iranian Cinema, Volume 3*, 14–15.

48. Susan Sontag's "The Decay of Cinema" was emblematic of this kind of eulogistic essay. Written around the time of cinema's centennial, the essay mourns the chemical decay of celluloid film. Sontag laments that the movies—once a "bigger-than-you" experience—have become a "variety of habit-forming home entertainments." For more, see Sontag, "The Decay of Cinema," *New York Times*, February 25, 1996; and Newman, *Video Revolutions*, 88–94.

49. Shamim Mostaghimi, "Az sinemā beh vide'o: faryād-hā va najvā-hā" [From cinema to video: Shouts and whispers], *Andisheh puyā* 22 (December 2014): 97.

50. Blake Atwood, *Reform Cinema in Iran: Film and Political Change in the Islamic Republic* (New York: Columbia University Press, 2016).

51. Negar Mottahedeh, *Displaced Allegories: Post-Revolutionary Iranian Cinema* (Durham, NC: Duke University Press, 2008); Naficy, *A Social History of Iranian Cinema, Volume 3*.

52. This is not to say that all such Iranians participated in the circulation and consumption of videocassettes or that other demographics did not have access to video. I encountered several people who claimed that they did not have much experience

with videocassettes during this period for a variety of reasons—strict parents, a lack of interest, or inaccessibility. Nevertheless, anecdotal evidence suggests that home video was a widespread phenomenon in the 1980s and 1990s.

53. Lobato and Thomas, *The Informal Media Economy*, 13.

54. Kowsar Avini, "Dokhtar-e shahid-e Avini: shakhsiyyat-e pedaram rā tekkeh-tekkeh kardeh-and" [Shahid Avini's daughter: They have fragmented my father's identity], *Tarikh Irani*, last modified April 9, 2013, http://tarikhirani.ir/fa/news /3096/.

55. For more about Morteza Avini's life, art, and career, see Roxanne Varzi, *Warring Souls: Youth, Media, and Martyrdom in Post-Revolution Iran* (Durham, NC: Duke University Press, 2006), 76–105; and Agnès Devictor, "Shahid Morteza Avini, Cinéaste et Martyr," *La pensée de midi* (February 2009): 54–60.

56. Avini, "Shahid Avini's Daughter."

57. For more on the formal and informal distribution of midcentury Iranian cinema, see Laura Fish, "Arisen from the Grave: Collecting and Distributing Midcentury Iranian Popular Cinema," Ph.D. diss., University of Texas at Austin, 2019.

58. James C. Scott, *Domination and the Arts of Resistance* (New Haven, CT: Yale University Press, 1990).

59. See also James C. Scott, *Weapons of the Weak: Everyday Forms of Peasant Resistance* (New Haven, CT: Yale University Press, 1985).

60. Susan Leigh Star, "Ethnography of Infrastructure," *American Behavioral Scientist* 43, no. 3 (1999): 381.

61. Changiz Pahlevan, "Negāhi beh mowqeʻiyaat-e yek rasāneh-ye khatarnāk" [A look at the place of a dangerous medium], *Māhnāmeh-ye Kelk* 1, no. 3 (May 1990): 106–110.

62. Abbas Kazemi, "Ezhdehāʾi keh tarbiyyat shod: masraf-e videʾo va zendegi-ye ruzmarreh-ye daheh-ye shast" [A dragon tamed: Consuming video and daily life in the 1980s], *Andisheh puyā* 22 (December 2014): 94.

63. Ghazian, "The Crisis in the Iranian Film Industry," 77–85.

64. Hamid Naficy, "Islamizing Film Culture in Iran: A Post-Khatami Update," in *New Iranian Cinema*, ed. Richard Tapper (London: I. B. Tauris, 2002): 26–65.

65. For more on this period, see Atwood, *Reform Cinema in Iran*; Naficy, "Islamizing Film Culture in Iran," 26–65; and Sussan Siavoshi, "Cultural Policies and the Islamic Republic: Cinema and Book Publication," *International Journal for Middle East Studies* 28, no. 4 (November 1997): 509–530.

66. On Hashemi Rafsanjani's liberalization polices, see Aliakbar Jafari, "Two Tales of a City: An Exploratory Study of Cultural Consumption among Iranian Youth," *Iranian Studies* 40, no. 3 (Spring 2007): 367–383.

67. Ezzati, *Video and Visual Media*, 37.

68. See, for example, "Āyā towlid va taksir-e navār-e vide'o beh sarfeh va salāh-e mā khāhad bud?" [Will the production and duplication of videocassettes be worthwhile and to our advantage?], *Ettelā'āt* (Tehran), November 23, 1993. According to reports at the time, the MCIG requested seven billion rials ($4.1 million) from the government in order to prepare for the legalization of video. For more, see "Āzādi-ye vide'o beh cheh mafhumi ast?" [What does the legalization of video mean?], *Sorush* 358, August 21, 1993, 47.

69. "Darbāreh-ye mā" [About us], *Mo'asseseh-ye rasāneh-hā-ye tasviri* [Visual Media Institute], accessed June 26, 2017, http://rasaneha.org/%D8%AF%D8%B1%D8% A8%D8%A7%D8%B1%D9%87-%D9%85%D8%A7.

70. "Ettelā'iyeh: Avvalin jashnvāreh-ye vide'uii-ye sureh" [Announcement: The first Sureh Video Festival], *Sorush* 664, October 2, 1994, 66.

71. "Siyāsat-e pardākht-e sahmiyeh beh tahiyeh konandegān-e film-hā-ye vide'oi e'lām shod" [A policy subsidizing video movie producers was announced], *Ettelā'āt* (Tehran), October 4, 1993.

72. Star, "Ethnography of Infrastructure," 382.

73. "Film-hā'i keh dar bāzār-e vide'oi khāhim did" [What films will we see in the video market?], *Sorush* 660, September 4, 1993, 42–43.

74. "Āzādi-ye vide'o kāri khub ammā hanuz mobham" [The legalization of video is a good thing but still ambiguous], *Sorush* 663, September 25, 1994, 52–53.

75. "The Legalization of Video Is a Good Thing," 53.

76. "What Does the Legalization of Video Mean," 47.

77. See, for example, "Ekrān-e film-e '50 kilu ālbālu' motevaqqef mishavad" [Screening of the film "50 kilos of sour cherries" will be suspended], *BBC Persian*, last modified June 6, 2016, http://www.bbc.com/persian/arts/2016/06/160606 _l51_50kg_sour_cherries_cinema.

CHAPTER 2

1. Lisa Parks, "Around the Antenna Tree: The Politics of Infrastructural Visibility," *Flow*, March 5, 2010, https://www.flowjournal.org/2010/03/flow-favorites -around-the-antenna-tree-the-politics-of-infrastructural-visibilitylisa-parks-uc -santa-barbara/. See also Nicole Starosielski, *The Undersea Network* (Durham, NC: Duke University Press, 2015).

2. Brian Larkin, "The Poetics and Politics of Infrastructure," *Annual Review of Anthropology* 42 (2013): 336.

3. Cited in Larkin, "The Poetics and Politics of Infrastructure," 336.

4. Bruno Latour, *Pandora's Hope: Essays on the Reality of Science Studies* (Cambridge, MA: Harvard University Press, 1999), 23.

5. Larkin, "The Poetics and Politics of Infrastructure," 336.

6. See Lily U. Nguyen, "Infrastructural Action in Vietnam: Inverting the Techno-Politics of Hacking in the Global South," *New Media and Society* 18, no. 4 (2016): 637–652; and Joanne Randa Nucho, *Everyday Sectarianism in Urban Lebanon: Infrastructures, Public Services, and Power* (Princeton: Princeton University Press, 2017).

7. Parks, "Around the Antenna Tree."

8. See, for example, Lisa Parks, "'Stuff You Can Kick': Toward a Theory of Media Infrastructures," in *Between Humanities and the Digital*, ed. Patrik Svensson and David Theo Goldberg (Cambridge, MA: MIT Press, 2015), 355–373.

9. See Brian Larkin, *Signal and Noise: Infrastructure and Urban Culture in Nigeria* (Durham, NC: Duke University Press, 2008).

10. "Iranian Film Biz Revisited: Lotsa U.S. Cassettes, Picture Backlog," *Variety*, June 6, 1984, 79.

11. Ezzati, *Video and Visual Media*, 13.

12. Susan Leigh Star, "Ethnography of Infrastructure," *American Behavioral Scientist* 43, no. 3 (1999): 381; Parks, "Around the Antenna Tree."

13. See Parks, "Around the Antenna Tree"; and Starosielski, *The Undersea Network*.

14. AbdouMaliq Simone, "People as Infrastructure: Fragments in Johannesburg," *Public Culture* 16, no. 3 (Fall 2004): 407–429.

15. Historical examples like postal sorting show how integral the human body is to storing, sorting, and circulating information. For more, see Parks, "'Stuff You Can Kick.'"

16. Nazila Fathi, *The Lonely War: One Woman's Account of the Struggle for Modern Iran* (New York: Basic Books, 2014), 74.

17. Fathi, *The Lonely War*, 75.

18. Michael A Cusumano, Yiorgos Mylonadis, and Richard S. Rosenbloom, "Strategic Maneuvering and Mass-Market Dynamics: The Triumph of VHS over Beta," *Business History Review* 55, no. 1 (Spring 1992): 54.

19. David Edgerton, *The Shock of the Old: Technology and Global History since 1900* (Oxford: Oxford University Press, 2011), 39.

20. "Iranian Film Biz Revisited," 79.

21. "Iranian Film Biz Revisited," 79.

22. Ravi Sundaram, *Pirate Modernity: Delhi's Media Urbanism* (London: Routledge, 2010), 97.

23. Sundaram, *Pirate Modernity*, 97.

24. Sundaram, 97.

25. Hilderbrand, *Inherent Vice*.

26. See, for example, Pang, *Creativity and Its Discontents*.

27. Mahsa Salamati, "Transnational Film Circulation in the Iranian Context."

28. Brian Larkin, "Degraded Images, Distorted Sounds: Nigerian Video and the Infrastructure of Piracy," *Public Culture* 16, no. 2 (Spring 2004): 309.

29. Hilderbrand, *Inherent Vice*, 3.

30. Hilderbrand, 163.

31. Hilderbrand, 176.

32. Pang, *Creativity and Its Discontents*, 198.

33. Star, "Ethnography of Infrastructure," 380.

34. Hilderbrand, *Inherent Vice*, 176.

35. Media Farzin, "The Video People," *Bidoun*, accessed December 15, 2020, https://bidoun.org/articles/the-video-people.

36. For a popular representation of this phenomenon, see Marjane Satrapi, *The Complete Persepolis* (New York: Pantheon, 2000).

37. Larkin, *Signal and Noise*, 6.

38. See, for example, Tom Glynn, *Reading Publics: New York City's Public Libraries, 1754–1911* (New York: Fordham University Press, 2017); and Tom Standage, *The Victorian Internet: The Remarkable Story of the Telegraph and the Nineteenth Century's On-line Pioneers*, 2nd ed. (New York: Bloomsbury USA, 2014).

39. For a rigorous historical account of the intersection of the Iranian film industry and cosmopolitanism before the revolution, see Golbarg Rekabtalaei, *Iranian Cosmopolitanism: A Cinematic History* (Cambridge: Cambridge University Press, 2018).

40. "Vorud-e navār-e khām-e video mamnu' shod" [The entry of blank tapes has been banned], *Film* (March 1986): 15.

41. In her memoir *Lipstick Jihad*, for example, journalist Azadeh Moaveni recalls the customs lines at Mehrabad Airport during her first trip back to Iran in 1999. With the latest issue of *Vogue* carefully tucked in her pants, she watched while the customs official ruffled through her bags and ultimately confiscated a mousepad depicting Botticelli's "The Birth of Venus," which he deemed inappropriate. For more, see Azadeh Moaveni, *Lipstick Jihad: A Memoir of Growing up Iranian in America and American in Iran* (New York: Public Affairs, 2005).

42. "Iranian Film Biz Revisited," 79.

43. "Mideast Entertainment Void Prompts Homevideo Bonanza," *Variety*, May 13, 1981, 370.

44. M. A. Khan, "Vidcassette Smuggling Major Headache to Pakistan Exhibs," *Variety*, May 28, 1980, 42.

45. Azar Nafisi, *Reading Lolita in Tehran: A Memoir in Books* (New York: Random House, 2003), 232.

46. Kazemi, "A Dragon Tamed," 95.

47. Kazemi, 94.

48. Fathi, *The Lonely War*, 73.

49. Mostaghimi, "From Cinema to Video," 97.

50. See Tung-Hui Hu, *A Prehistory of the Cloud* (Cambridge, MA: MIT Press, 2016); and Starosielski, *The Undersea Network*.

51. Dana Stevens, "The Cab Ride as Artistic Rebellion," *Slate*, October 2, 2015, https://slate.com/culture/2015/10/jafar-panahis-film-taxi-reviewed.html.

52. Parks, "Around the Antenna Tree."

53. See, for example, Richard Brody, "Jafar Panahi's Remarkable 'Taxi,'" *New Yorker*, October 13, 2015, https://www.newyorker.com/culture/richard-brody/jafar-panahis-remarkable-taxi; and Jonathan Romney, "Taxi Tehran Review—Jafar Panahi's Joyous Ride," *The Guardian*, last modified November 1, 2015, https://www.theguardian.com/film/2015/nov/01/taxi-tehran-review-admirable-jafar-panahi-takes-to-the-streets.

54. Romney, "Taxi Tehran Review."

55. Lobato, *Shadow Economies of Cinema*, 69–94.

CHAPTER 3

1. My use of "aspirational work" here is inspired by Brooke Erin Duffy's study on aspirational digital labor in the United States. She builds off "the much-venerated ideal of *getting paid to do what you love*." Her study ties aspirational work to financial mobility, as social media influencers and vloggers chase the dream of marrying their passions with their profession. Aspirants engage in laborious work online for free in the hope of being paid to pursue their interests at some point in the future. The case of underground video dealing in Iran expands this definition of aspirational media labor to include social and cultural capital, in addition to financial gain. For more, see Brooke Erin Duffy, *(Not) Getting Paid to Do What You Love: Gender, Social Media, and Aspirational Work* (New Haven: Yale University Press, 2017).

2. Herbert, *Videoland*, 72.

3. For more on the new economy and cultural work, see Angela McRobbie, *Be Creative: Making a Living in the New Culture Industries* (Cambridge: Polity Press, 2016).

4. For more on informal circuits of film distribution in Iran, see Kaveh Askari, "An Afterlife for Junk Prints: Serials and Other 'Classics' in Late-1920s Tehran," in *Silent Cinema and the Politics of Space*, ed. Jennifer M. Bean, Anupama P. Kapse, and Laura Evelyn Horak (Bloomington: Indiana University Press, 2014), 99–120; and Kaveh Askari, "Eastern Boys and Failed Heroes: Iranian Cinema in the World's Orbit," *Cinema Journal* 57, no. 3 (Spring 2018): 29–53.

5. Greenberg, *From Betamax to Blockbuster*, 68–72.

6. "Sargozasht-e vide'o dar Iran, qesmat-e avval" [The tale of video in Iran, part 1], *Sorush* 657, August 21, 1993, 32–33; Ezzati, *Video and Visual Media*, 13.

7. Ezzati, 13.

8. Ezzati, 18.

9. Ezzati, 14.

10. The trade *Tamāshā* (*Watching*), for example, includes only eight references to video in the 188 issues published between 1975 and 1979, mostly in the form of advertisements and translated articles.

11. For more on video and the rhetoric of revolution, see Newman, *Video Revolutions*, 36–37.

12. "Sargozasht-e vide'o dar Iran, qesmat-e dovom" [The tale of video in Iran, part 2], *Sorush* 659, September 4, 1993, 44.

13. Farzin, "The Video People."

14. Farzin, "The Video People."

15. Farzin, "The Video People."

16. Jacques Rancière, *Proletarian Nights: The Workers' Dream in Nineteenth-Century France*, trans. John Drury (New York: Verso, 2012).

17. Rancière, *Proletarian Nights*, 3–23.

18. Naficy, *A Social History of Iranian Cinema, Volume 1*, 202.

19. Naficy, 202.

20. Naficy, *A Social History of Iranian Cinema, Volume 2*, 158.

21. The mostly popular genres—including romance, crime, and action—that populated second- and third-class cinemas were intended to distract the masses. As Naficy claims, at this time "the government kept ticket prices low . . . to appease the potentially volatile lower-class publics." Naficy, *A Social History of Iranian Cinema, Volume 2*, 158.

22. *Sorkhpust*, literally "red skin," is a loaded racial term used in Persian to describe Native Americans. The title is likely a riff on the popular film motif "cowboys and Indians." I have translated it here as "Indians" because I believe it is historically accurate not because I think that "Indians" is the proper designation for Native Americans. On an entirely separate note, I would like to thank Babak Tabarraee for alerting me to this film.

23. "The Tale of Video in Iran, Part 2," 44.

24. "The Tale of Video in Iran, Part 2," 44.

25. Kathleen M. Millar, "The Precarious Present: Wageless Labor and Disrupted Life in Rio de Jeneiro, Brazil," *Cultural Anthropology* 29, no. 1 (2014): 35.

26. These assumptions about the "creative class" were the basis for the earliest theorizations of creative labor. See, for example, Richard Florida, *The Rise of the Creative Class and How It's Transforming Work, Leisure, Community and Everyday Life* (New York: Basic Books, 2002).

27. Early efforts to theorize creative labor gave credence to the myths of Hollywood and understood creative workers like directors, actors, and screenwriters simply as artisans who led a luxurious, often bohemian, lifestyle. Over the last decade, however, scholars have sought to make visible entire workforces within creative industries whose lives looked very different than those of the bohemian artisans at the core of early theorizations of creative work. These so-called below-the-line workers—including make-up artists, set designers, grips, and lighting technicians—generally do not receive credit for creating anything, but their work, nevertheless, requires great ingenuity. In these accounts, creative work is not simply the domain of directors and actors but rather of a wide-range of agents. See, for example, John Thornton Caldwell, *Production Culture: Industrial Reflexivity and Critical Practice in Film and Television* (Durham, NC: Duke University Press, 2008); Vicki Mayer, *Below the Line: Producers and Production Studies in the New Television Economy* (Durham, NC: Duke University Press, 2011); and Erin Hill, *Never Done: A History of Women's Work in Media Production* (New Brunswick, NJ: Rutgers University Press, 2016).

28. Pierre Bourdieu, *Distinction: A Social Critique of the Judgement of Taste*, trans. Richard Nice (Cambridge, MA: Harvard University Press, 1984), 359.

29. Keith Negus, "The Work of Cultural Intermediaries and the Enduring Distance between Production and Consumption," *Cultural Studies* 16, no. 4 (2002): 505.

30. Bourdieu, *Distinction*, 359.

31. Negus, "The Work of Cultural Intermediaries," 505.

32. Lobato, *Shadow Economies of Cinema*, 70.

33. Lobato, 61–94.

34. Lobato, 78–80.

35. Ramon Lobato, "The Six Faces of Piracy: Global Media Distribution from Below," in *The Business of Entertainment, Volume 1*, ed. R. C. Sickels (Westport, CT: Greenwood Press, 2008), 27.

36. See, for example, Jimmy Draper, "Theorizing Creative Agency through 'Discerned Savvy': A Tool for the Critical Study of Media Industries," *Media, Culture, and Society* 36, no. 8 (2014): 1118–1133; and Timothy Havens and Amanda D. Lotz, *Understanding Media Industries* (New York: Oxford University Press, 2015), 13.

37. Farzin, "The Video People."

38. Farzin, "The Video People."

39. Farzin, "The Video People."

40. Herbert, *Videoland*, 2–3.

41. For a comprehensive account of film genres as discursive categories, see Rick Altman, *Film/Genre* (London: British Film Institute, 1999).

42 "Vorud-e navār-e khām-e vide'o mamnu' shod" [The entry of blank tapes has been banned], *Film* 34 (1986), 10.

43. Farzin, "The Video People."

44. See, for example, Hamid Naficy, *The Making of Exile Cultures: Iranian Television in Los Angeles* (Minneapolis: University of Minnesota Press, 1993).

45. Fardin Alikhah, "The Politics of Satellite Television in Iran," in *Media, Culture, and Society: Living with Globalization and the Islamic State*, ed. Mehdi Semati (London: Routledge, 2008), 94–110.

46. Media scholars have begun to take *fandubbing* and *fansubbing* seriously. These studies show how the popular practice of fans adding their own subtitles or dubbing movies challenges understandings of transnationalism, authorship, and formal industry. The studies almost always attribute the rise of this kind of fan labor to digital technologies. The case of video dealers in Iran shows how analog video technology served as an important precedent to these digital practices. In this scheme, we can also understand some video dealers as traditional fans, who were often creative and productive in their own engagement with media. See, for example, Hye-Kyung Lee, "Participatory Media Fandom: A Case Study of Anime Fansubbing," *Media, Culture & Society* 33, no. 8 (2011): 1131–1147; and Matt Hills, "Transnational Cult and/as Neoliberalism: The Liminal Economies of Anime Fansubbers," *Transnational Cinemas* 8, no. 1 (2016): 80–94.

47. "Sedā-ye to khub ast: behtarin-hā-ye tārikh-e dubleh Iran" [Your voice is good: The best in the history of dubbing in Iran], *Ketāb-e sāl-e sinemā-ye Iran, 1384* [The yearbook of Iranian cinema, 2005/2006] 15 (February 2006): 120.

48. Shapur Azimi, "Sedā-ye āshenā: negāhi be dubleh-ye *Bar bād rafteh*" [The familiar voice: A look at the dubbing of *Gone with the Wind*], *Film*, no. 404 (2010): 32.

49. Babak Tabarraee, "*Gone with the Wind* in Iran: The Iranian Dub Industry and the Re-authorization of Classical Hollywood Cinema," YouTube video, 39:38, January 18, 2018, https://www.youtube.com/watch?v=j7bA7_KCWXs.

CHAPTER 4

1. Lynn Spigel, *Make Room for TV: Television and the Family Ideal in Postwar America* (Chicago: University of Chicago Press, 1992), 4–8.

2. See, for example, Benson-Allott, *Killer Tapes and Shattered Screens*; Gray, *Video Playtime*; and Klinger, *Beyond the Multiplex*.

3. Klinger, 11.

4. Benson-Allott, *Killer Tapes and Shattered Screens*, 24.

5. Klinger, *Beyond the Multiplex*, 17.

6. Kazemi, "A Dragon Tamed," 94.

7. Starosielski, *The Underground Network*, 82.

8. "Porsesh az dāneshjuyān: Cherā pā-ye vide'o mi-neshanid va az ān cheh estefādeh-hā'i mi-konid?" [Asking university students: Why do you watch video and what do you do with it?], *Gardun* (September–October 1993): 24.

9. Kazemi, "A Dragon Tamed," 94.

10. Mostaghimi, "From Cinema to Video," 97.

11. Caetlin Benson-Allott, *Remote Control* (New York: Bloomsbury Academic, 2015), xviii.

12. In fact, it was common enough that Iran named its own video-sharing platform "Aparat," a reference to the projectors that had supported early home viewing experiences. Because YouTube was (and still is) blocked in Iran, Aparat found quick success, which it has maintained since the site began operating in 2011. According to agencies that collect web traffic statistics, Aparat currently is the second most frequented website in Iran. A name like Aparat shows how media infrastructures map on to one another, as the word for "home projectors" also became emblematic of today's large-scale video sharing. "aparat.com Traffic Statistics," Alexa, https://www.alexa.com/siteinfo/aparat.com#trafficstats.

13. Kazemi, "A Dragon Tamed," 94.

14. Kazemi, 94.

15. For more on the transformations of domestic space throughout the nineteenth and twentieth centuries, see Pamela Karimi, "Gendered Spaces and Bodies Out of

Place," *Domesticity and Consumer Culture in Iran: Interior Revolutions of the Modern Era* (New York: Routledge, 2011), 143–162.

16. See, for example, Naficy, *A Social History of Iranian Cinema, Volume 1*; Mottahedeh, *Displaced Allegories*; and Saeed Zeydabadi-Nejad, *The Politics of Iranian Cinema: Film and Society in the Islamic Republic* (London: Routledge, 2009).

17. Naficy, *A Social History of Iranian Cinema, Volume 1*.

18. Mostaghimi, "From Cinema to Video," 97.

19. Mostaghimi, 97.

20. Mostaghimi, 98.

21. Jean Baudrillard, *The System of Objects*, trans. James Benedict (London and New York: Verso, 2005).

22. Christian Metz, *The Imaginary Signifier: Psychoanalysis and the Cinema*, trans. Celia Britton (Bloomington: Indiana University Press, 1982), 57.

23. Walter Benjamin, "Unpacking My Library: A Talk about Book Collecting," in *Illuminations*, ed. Hannah Arendt, trans. Harry Zohn (New York: Schocken, 1978), 67.

24. Benjamin was famously obsessed with collectors precisely because they took commodities out of marketplace circulation and assigned a different kind of value to them. See Walter Benjamin, *The Arcades Project*, trans. Howard Eiland and Kevin McLaughlin (Cambridge, MA; London: Belknap Press of Harvard University Press, 1999).

25. See, for example, Kim Bjarkman, "To Have and to Hold: The Video Collector's Relationship with an Ethereal Medium," *Television and New Media* 5, no. 3 (August 2005): 217–246; and Charles Tashiro, "The Contradictions of Video Collecting," *Film Quarterly* 50, no. 2 (Winter 1996–1997): 11–18. Also, for this reason, critical media infrastructure studies, especially the kind of phenomenological approach advocated by Lisa Parks, proves more productive to understanding the kind of attachments that people formed to videocassettes in Iran, where collecting and ownership of videotapes were rare phenomena.

26. Hilderbrand, *Inherent Vice*, 197.

27. Hilderbrand, 183–190.

28. Shahriar Mandanipour, *Censoring an Iranian Love Story*, trans. Sara Khalil (New York: Vintage Books, 2009).

29. Varzi, *Warring Souls*, 116.

30. Tara Bahrampour, *To See and See Again: A Life in Iran and America* (Berkeley: University of California Press, 2000), 283.

31. Bahrampour, *To See and See Again*, 284.

32. Blake Atwood, "When the Sun Goes Down: Sex, Desire, and Cinema in Tehran," *Asian Cinema* 27, no. 2 (October 2016): 127–150.

33. Pardis Mahdavi, *Passionate Uprisings: Iran's Sexual Revolution* (Stanford, CA: Stanford University Press, 2009).

34. Linda Williams, *Screening Sex* (Durham, NC: Duke University Press, 2008), 6.

35. Shahram Khosravi, *Young and Defiant in Tehran* (Philadelphia: University of Pennsylvania Press, 2009), 156.

36. For more on *Titanic*'s global reception as it relates to the film's informal distribution, see Klinger, "Contraband Cinema."

37. For more on obscenity and cinema, see Lotte Hoek, *Cut-Pieces: Celluloid Obscenity and Popular Cinema in Bangladesh* (New York: Columbia University Press, 2013).

38. For more, see Blake Atwood, "The Little Devil Comes Home: Video, the State, and Amateur Cinema in Iran," *Film History* 30, no. 1 (Spring 2018): 138–167.

39. Ezzati, *Video and Visual Media*, 50–51.

40. See, for example, Peter Alilunas, *Smutty Little Movies: The Creation and Regulation of Adult Video* (Berkeley: University of California Press, 2016); Herbert, *Videoland*, 62–63; Greenberg, *From Betamax to Blockbuster*, 50–52; and Wasser, *Veni, Vidi, Video*, 92–95.

41. See, for example, Jane Juffer, *At Home with Pornography: Women, Sexuality, and Everyday Life* (New York: New York University Press, 1998); and Linda Williams, *Hard Core: Power, Pleasure, and the "Frenzy of the Visible"* (Berkeley: University of California Press, 1989).

42. Juffer, *At Home with Pornography*, 233.

43. Moaveni, *Lipstick Jihad*, 185.

44. Moaveni, 187.

45. Moaveni, 188.

46. Moaveni, 188.

47. Moaveni, 189.

48. Moaveni, 188.

49. Moaveni, 190.

50. As Hilderbrand notes, Jane Fonda's *Workout* (1982) video revealed a marketplace logic wherein early prerecorded content on video most often lent itself to private consumption. For Hilderbrand, "the self-consciousness many users would have felt about exercising in public aerobics classes may have paralleled the embarrassment many would have felt in adult theaters; thus body shame may have been central to home video's early success." As we can see, the global success of Fonda's workout videos shows how home video's associations with private subjectivities

existed in many places around the world, even if for varying reasons. For more, see Hilderbrand, *Inherent Vice*, 59–60.

51. Varzi, *Warring Souls*, 133.

52. Varzi, 138.

53. Spigel, *Make Room for TV*, 52.

54. See, for example, Mehdi Ganjavi, "A TV Show for the Rouhani Era: *Shahrzad* and the 1953 Coup d'etat in Iran," *Ajam Media Collective* (August 15, 2016), https:// ajammc.com/2016/08/15/shahrzad-1953-coup-in-iran/; and Khatereh Sheibani, "Shahrzad: Farhang-e mardomi, farhang-e rasāneh-i va bāznamā'i-ye jensiyyat dar 'asr-e dijitāl" [*Shahrzad*: Pop culture, media culture, and the representation of gender in the digital age], *Iran Namag: A Quarterly of Iranian Studies* 2, no. 2 (Summer 2017): 62–111.

55. Sheibani, "*Shahrzad*," 64.

56. Ghazaleh Haghdad Mofrad, "The Emergence of 'Home Shows': The Market and Politics of Television Entertainment in Iran," *Ajam Media Collective*, last modified November 14, 2014, https://ajammc.com/2014/11/17/iranian-home-shows/.

CHAPTER 5

1. Jennifer Gabrys, *Digital Rubbish: A Natural History of Electronics* (Ann Arbor: University of Michigan Press, 2011): 16.

2. Charles R. Acland, "Introduction" in *Residual Media*, ed. Charles R. Acland (Minneapolis: University of Minnesota Press, 2007), xix–xx.

3. See, for example, "Dastgāh-hā-ye paksh-e navār-e vide'o bel'akhareh beh tārikh payvastand" [Video players finally become history], *'Asr-e Iran*, last modified July 23, 2016, http://www.asriran.com/fa/news/481480/; and "Navār-e vide'o rasman mord" [The videocassette has officially died], *Bartarin-hā*, last modified July 23, 2016, http://www.bartarinha.ir/fa/news/357615/.

4. Marianne Hirsch and Leo Spitzer, "Testimonial Objects: Memory, Gender, and Transmission," *Poetics Today* 27, no. 2 (Summer 2006): 355.

5. Wendy Hui Kyong Chun, *Programmed Visions: Software and Memory* (Cambridge, MA: MIT Press, 2011), xii.

6. Jeff Scheible, "Video after Video Stores," *Canadian Journal of Film Studies* 23, no. 1 (Spring 2014): 54.

7. Scheible, "Video after Video Stores," 53.

8. For more on Attaran and the recent trend of absurdist popular cinema in Iran, including *Sperm Whale*, see Babak Tabarraee, "Rationalizing the Irrational: Reza Attaran's Popularity, Stardom, and the Recent Cycle of Iranian Absurd Films," *Iranian Studies* 51, no. 4 (2018): 613–632.

9. The fact that only Ayatollah Khomeini's picture is hanging in the station (and not Ayatollah Khamenei, the current Supreme Leader in Iran) helps us date this scene in the film. We know that it must be taking place before 1989 when Khomeini died and was replaced by Khamenei. The video ban would therefore have been in full effect.

10. Tabarraee, "Rationalizing the Irrational," 625.

11. "Naqd va barrasi-ye 'Nahang-e 'anbar-e 2: selekshen-e Roya'" [A review of "Sperm whale 2: Roya's selection"], *Naqd-e farsi*, last modified June 2, 2017, http://www.naghdefarsi.com/iran-movie-review/18462-nahang-anbar.html; Sina Tahmuresi, "Negāhi beh film-e 'Nahang-e 'anbar-e 2'" [A look at "Sperm whale 2"], *Zoomg*, last modified May 29, 2017, http://www.zoomg.ir/2017/5/21/161561/movie-nahang-e-anbar-introduction/.

12. Tabarraee, "Rationalizing the Irrational," 614–615.

13. See, for example, Nahid Siamdoust, *Soundtrack of the Revolution: The Politics of Music in Iran* (Stanford: Stanford University Press, 2017).

14. See, for example, "Nazar-e Mahnaz Afshar darbāreh-ye *Nahang-e 'anbar* va nostālzhi-ye daheh-ye shast" [Mahnaz Afshar's opinion about *Sperm Whale* and the 1980s generation's nostalgia], *Nasim News Agency*, last modified August 4, 2015, http://old.nasimonline.ir/Content/Detail/1017748/; and Saeed Morovvati, "Nostālzhi-ye posht-e mu va shalvār khomrehi" [The nostalgia behind the hair and the baggy pants], *Hamshahri*, last modified February 9, 2019, http://newspaper.hamshahri.org/id/47172/.

15. Fred Davis, "Nostalgia, Identity, and the Current Nostalgia Wave," *Journal of Popular Culture* 11, no. 2 (Fall 1977): 414–424.

16. Boym, *The Future of Nostalgia*, 7.

17. For example, Linda Hutcheon and Mario J. Valdes write that "nostalgia is not something you 'perceive' in an object; it is what you 'feel' when two different temporal moments, past and present, come together for you and often, carry considerable emotional weight" (23). For more, see Hutcheon and Valdes, "Irony, Nostalgia, and the Postmodern: A Dialogue," *Poligraphias* 3 (2000): 18–41. On the other hand, Svetlana Boym famously declared that "nostalgia is about the relationship between individual biography and the biography of groups or nations, between personal and collective memory." Boym, *The Future of Nostalgia*, xvi.

18. Manuel Menke, "Seeking Comfort in Past Media: Modeling Media Nostalgia as a Way of Coping with Media Change," *International Journal of Communication*, no. 11 (2017): 626–646.

19. See, for example, "Nejāt-e 'Nahang-e 'anbar' bā bāzigushi-ye Attaran / Tekrār-e filmfārsi-hā-ye bi-hoviyyat" [The redemption of "Sperm Whale" with Attaran's mischief/The repetition of nondescript filmfārsi movies], *Farhang News*, last modified July 4, 2015, http://www.farhangnews.ir/content/130946.

20. Tabarraee, "Rationalizing the Irrational," 615.

21. "*Nahang-e ʿanbar* vāqaʿan arzesh-e didār dārad" [*Sperm Whale* is truly worth seeing], *Persianv*, last modified June 1, 2015, http://persianv.com/cinema/%D9%81% DB%8C%D9%84%D9%85-%D9%86%D9%87%D9%86%DA%AF-%D8 %B9%D9%86%D8%A8%D8%B1-%D9%88%D8%A7%D9%82%D8%B9%D8%A7 -%D8%A7%D8%B1%D8%B2%D8%B4-%D8%AF%DB%8C%D8%AF%D9%86 -%D8%AF%D8%A7%D8%B1%D8%AF-%D8%AA%D8%B5.html.

22. Pedram Partovi, *Popular Iranian Cinema before the Revolution: Family and Nation in Filmfārsī* (New York: Routledge, 2017), 70–71.

23. Tabarraee, "Rationalizing the Irrational," 620–621.

24. Tabarraee, 623.

25. "Nahang-e ʿanbar: ehyā-ye film-hā-ye kābārehi" [Sperm whale: The 1980s generation's nostalgia for the revival of cabaret films], *Cinema Press*, last modified December 14, 2017, http://cinemapress.ir/news/91669.

26. "Nahang-e ʿanbar: tekrār-e filmfārsi-hā-ye bi-hoviyyat" [Sperm whale: The repetition of nondescript filmfārsis], *Jām News*, last modified July 4, 2015, http:// www.jamnews.com/detail/News/501515.

27. Tabarraee, "Rationalizing the Irrational," 626.

28. "Iranian Film Biz Revisited," 91.

29. Esmaeil Mohindoost, *Ru dar ru bā Asghar Farhadi* [In conversation with Asghar Farhadi] (Tehran: Rownaq, 2016), 20.

30. Mihandoost, *In Conversation with Asghar Farhadi*, 14.

31. Mihandoost, 28.

32. Mihandoost, 28.

33. "Mani Haghighi: mikhāham bā kārgardāni-ye '50 kilu ālbālu' sinemā-ye tejāri rā tajrobeh konam" [Mani Haghighi: By directing "50 kilos of sour cherries, I want to experiment with commercial cinema], *Salām Sinamā*, last modified August 29, 2015, https://www.salamcinama.ir/news/2659/.

34. "Jadval-e forush: 1395" [Box office sales: 1395], https://www.cinematicket .org/?p=boxoffice.

35. "Ekrān-e film-e '50 kilu ālbālu' motevaqqef mishavad" [Screening of the film "50 kilos of sour cherries" will be suspended], *BBC Persian*, last modified June 6, 2016, http://www.bbc.com/persian/arts/2016/06/160606_l51_50kg_sour_cherries _cinema.

36. "Jadidtarin jelveh filmfārsi dar 'Panjāh kilo ālbālu'" [The newest sparkle of filmfārsi in "50 kilos of sour cherries"], *Cinema Press*, last modified April 17, 2016, http://www.iranchamber.com/calendar/converter/iranian_calendar_converter .php.

37. "Porforushtarin film-hā-ye sinemā-ye Iran dar sāl-e 1391 hamrāh bā 'aks" [The best-selling movies in Iranian cinema in 1391 along with pictures], *Bizhan Zadmehr*, February 24, 2013, http://bijanzadmehr.blogfa.com/post/59/.

38. "Film-e 'Zedd-e goluleh,'" [The film "Bulletproof"], *Bāshgāh-e film-e Tehran*, last modified September 2, 2011, http://filmclub.vcp.ir/142558-%D9%81%DB%8C%D9%84%D9%85-%D8%B6%D8%AF-%DA%AF%D9%84%D9%88%D9%84%D9%87.

39. "Mostafa Kiyaee: Filsuf nistam va sho'ār nemidaham/Fajr barā-ye man vitrin ast" [Mostafa Kiyaee: I am not a philosopher nor a fan of slogans/Fajr is a showcase for me], *Islamic Azad University News Agency*, February 17, 2016, http:// https://ana.press/fa/news/63/65872/.

40. Over the last three decades, media historians have found renewed purpose in combating the idea that digital media is "new" media. Purposefully disrupting easy historical narratives, they have often focused on comparative newness, reading historical moments of innovation against one another. In *Always Already New*, Lisa Gitelman emphasizes "the novelty years" of a medium as central to how we conceive of "the course of media history." I do not disagree with the idea that the "novelty years" reveal how media have "inscribed" themselves, as Gitelman claims, but I am also interested in theorizing media use beyond its initial diffusion and even into its afterlife. Gitelman, *Always Already New: Media, History, and the Data of Culture* (Cambridge, MA: MIT Press, 2006), 1.

41. For more on media collecting, see Kate Egan, "The Celebration of a 'Proper Product': Exploring the Residual Collectible through the 'Video Nasty,'" in *Residual Media: Residual Technologies and Culture*, ed. Charles Acland (Minneapolis: University of Minnesota Press, 2007), 200–221; Klinger, *Beyond the Multiplex*, 54–90; Gregory Steirer, "The Personal Media Collection in an Era of Connected Viewing" in *Connected Viewing: Selling, Streaming and Sharing Media in the Digital Age*, ed. Jennifer Holt and Kevin Sanson (New York: Routledge, 2014), 79–96; and Tashiro, "The Contradiction of Video Collection."

42. Farzin, "The Video People."

43. Farzin, "The Video People."

44. Babak Tabarraee, *Kālt* [Cult] (Tehran: Enteshārāt-e Qoqnus, 2016), 66.

CODA

1. Amir Ezzati, "Hekāyat-e enteshār-e ketāb bedun-e nām-e mo'allef" [The story of publishing a book without the name of the author], *Video va rasāneh-hā-ye tasviri* (Tehran: Cinema Museum, 2004), n.p.

2. F. Parsa, "A Letter from Tehran," trans. Dorna Khazeni, *Film-Philosophy* 7, no. 12 (June 2003), http://www.film-philosophy.com/vol7-2003/n12parsa.

3. Ezzati, "The Story of Publishing a Book without the Name of the Author," n.p.

4. Ezzati, n.p.

5. Ezzati, n.p.

6. Atwood, *Reform Cinema in Iran.*

7. Ezzati, *Video and Visual Media*, 58.

8. For more on the development of the internet in Iran, see Niki Akhavan, *Electronic Iran: The Cultural Politics of an Online Evolution* (New Brunswick, NJ: Rutgers University Press, 2013).

9. Mary Carruthers, *The Book of Memory: A Study of Memory in Medieval Culture* (Cambridge: Cambridge University Press, 2008), 1.

BIBLIOGRAPHY

Abrahamian, Ervand. *A History of Modern Iran*. Cambridge: Cambridge University Press, 2008.

Acland, Charles R., ed. *Residual Media*. Minneapolis: University of Minnesota Press, 2007.

Akhavan, Niki. *Electronic Iran: The Cultural Politics of an Online Evolution*. New Brunswick, NJ: Rutgers University Press, 2013.

Alikhah, Fardin. "The Politics of Satellite Television in Iran." In *Media, Culture, and Society: Living with Globalization and the Islamic State*, edited by Mehdi Semati, 94–110. London: Routledge, 2008.

Alilunas, Peter. *Smutty Little Movies: The Creation and Regulation of Adult Video* (Berkeley: University of California Press, 2016).

Altman, Rick. *Film/Genre*. London: British Film Institute, 1999.

"Āmār-e vaz'iyat-e sinemā, televiziyun va vide'o dar 42 keshvar-e jahān" [Statistics on the status of cinema, television, and video in 42 countries around the world]. *Film*. January 1993.

"aparat.com Traffic Statistics." Alexa. Accessed June 30, 2018. https://www.alexa.com/siteinfo/aparat.com#trafficstats.

Askari, Kaveh. "An Afterlife for Junk Prints: Serials and Other Classics in Late-1920s Tehran." In *Silent Cinema and the Politics of Space*, edited by Jennifer M. Bean, Anupama P. Kapse, and Laura Evelyn Horak, 99–120. Bloomington: Indiana University Press, 2014.

Askari, Kaveh. "Eastern Boys and Failed Heroes: Iranian Cinema in the World's Orbit." *Cinema Journal* 57, no. 3 (Spring 2018): 29–53.

Athique, Adrian. "The Global Dynamics of Indian Media Piracy: Export Markets, Playback Media and the Informal Economy." *Media, Culture & Society* 30, no. 5 (2008): 699–717.

Atwood, Blake. "The Little Devil Comes Home: Video, the State, and Amateur Cinema in Iran." *Film History* 30, no. 1 (Spring 2018): 138–167.

Atwood, Blake. *Reform Cinema in Iran: Film and Political Change in the Islamic Republic*. New York: Columbia University Press, 2016.

Atwood, Blake. "When the Sun Goes Down: Sex, Desire, and Cinema in Tehran." *Asian Cinema* 27, no. 2 (October 2016): 127–150.

Avini, Kowsar. "Dokhtar-e shahid-Avini: shakhsiyyat-e pedaram rā tekkeh-tekkeh kardeh-and" [Shahid Avini's daughter: They have fragmented my father's identity]. *Farāru.* Last modified April 9, 2013. https://fararu.com/fa/news/146016.

"Āyā towlid va taksir-e navār-e vide'o beh sarfeh va salāh-e mā khāhad bud?" [Will the production and duplication of videocassettes be worthwhile and to our advantage?]. *Ettelāʿāt.* November 23, 1993.

"Āzādi-ye vide'o beh cheh mafhumi ast?" [What does the legalization of video mean?]. *Sorush.* August 21, 1993.

"Āzādi-ye vide'o kāri khub ammā hanuz mobham" [The legalization of video is a good thing but still ambiguous]. *Sorush.* September 25, 1994.

Azimi, Shapur. "Sedā-ye āshenā: negāhi be dubleh-ye *bar bād rafteh.*" [The familiar voice: A look at the dubbing of *Gone with the Wind*]. *Film,* no. 404 (January 2010): 32–33.

Bahrampour, Tara. *To See and See Again: A Life in Iran and America.* Berkeley: University of California Press, 2000.

Bajoghli, Narges. *Iran Reframed: Anxieties of Power in the Islamic Republic.* Stanford, CA: Stanford University Press, 2019.

Barad, Karen. *Meeting the Universe Halfway: Quantum Physics and the Entanglement of Matter and Meaning.* Durham, NC: Duke University Press, 2007.

Baudrillard, Jean. *The System of Objects.* Translated by James Benedict. London and New York: Verso, 2005.

Benjamin, Walter. *The Arcades Project.* Translated by Howard Eiland and Kevin McLaughlin. Cambridge, MA; London: Belknap Press of Harvard University Press, 1999.

Benjamin, Walter. "Unpacking My Library: A Talk about Book Collecting." In *Illuminations,* edited by Hannah Arendt, translated by Harry Zohn, 59–68. New York: Schocken, 1978.

Benson-Allott, Caetlin. *Killer Tapes and Shattered Screens: Video Spectatorship from VHS to File Sharing.* Berkeley: University of California Press, 2013.

Benson-Allott, Caetlin. *Remote Control.* New York: Bloomsbury Academic, 2015.

Benson-Allott, Caetlin. *The Stuff of Spectatorship: Material Cultures of Film and Television* (Berkeley: University of California Press, 2021).

Bjarkman, Kim. "To Have and to Hold: The Video Collector's Relationship with an Ethereal Medium." *Television and New Media* 5, no. 3 (August 2005): 217–246.

Boym, Svetlana. *The Future of Nostalgia.* New York: Basic Books, 2001.

Bourdieu, Pierre. *Distinction: A Social Critique of the Judgement of Taste*. Translated by Richard Nice. Cambridge, MA: Harvard University Press, 1984.

Brody, Richard. "Jafar Panahi's Remarkable 'Taxi.'" *New Yorker*. October 13, 2015. https://www.newyorker.com/culture/richard-brody/jafar-panahis-remarkable -taxi.

Caldwell, John Thornton. *Production Culture: Industrial Reflexivity and Critical Practice in Film and Television*. Durham, NC: Duke University Press, 2008.

Carruthers, Mary. *The Book of Memory: A Study of Memory in Medieval Culture*. Cambridge: Cambridge University Press, 2008.

Chelkowski, Peter, and Hamid Dabashi, *Staging a Revolution: The Art of Persuasion in the Islamic Republic of Iran*. New York: New York University Press, 1999.

Chun, Wendy Hui Kyong. *Programmed Visions: Software and Memory*. Cambridge, MA: MIT Press, 2011.

Cusumano, Michael A., Yiorgos Mylonadis, and Richard S. Rosenbloom. "Strategic Maneuvering and Mass-Market Dynamics: The Triumph of VHS over Beta." *Business History Review* 55, no. 1 (Spring 1992): 51–59.

"Darbāreh-ye mā" [About us]. *Mo'asseseh-ye rasāneh-hā-ye tasviri* [Visual Media Institute]. Accessed March 3, 2018. http://rasaneha.org/%D8%AF%D8%B1%D8 %A8%D8%A7%D8%B1%D9%87-%D9%85%D8%A7.

"Dastgāh-hā-ye paksh-e navār-e vide'o bel'akhareh beh tārikh payvastand" [Video players finally become history]. *'Asr-e Iran*. Last modified July 23, 2016. http://www.asriran.com/fa/news/481480/.

Davis, Fred. "Nostalgia, Identity, and the Current Nostalgia Wave." *Journal of Popular Culture* 11, no. 2 (Fall 1977): 414–424.

Decherney, Peter. *Hollywood's Copyright Wars: From Edison to the Internet*. New York: Columbia University Press, 2012.

Devictor, Agnès. "Shahid Morteza Avini, Cinéaste et Martyr." *La pensée de midi* (February 2009): 54–60.

Dostoyevsky, Fyodor. *Notes from the Underground*. Translated by Richard Pevear and Larissa Volokhonsky. New York: Vintage, 1994.

Douglas, Susan. "Writing from the Archive: Creating Your Own." *Communication Review* 13, no. 1 (2010): 5–14.

Draper, Jimmy. "Theorizing Creative Agency through 'Discerned Savvy': A Tool for the Critical Study of Media Industries." *Media, Culture, and Society* 36, no. 8 (2014): 1118–1133.

Duffy, Brooke Erin. *(Not) Getting Paid to Do What You Love: Gender, Social Media, and Aspirational Work*. New Haven: Yale University Press, 2017.

Edgerton, David. *The Shock of the Old: Technology and Global History since 1900.* Oxford: Oxford University Press, 2011.

Edwards, Paul N. "Infrastructure and Modernity: Force, Time, and Organization in the History of Sociotechnical Systems." In *Modernity and Technology*, edited by Thomas J. Misa, Philip Brey, and Andrew Freenberg, 185–225. Cambridge, MA: MIT Press, 2003.

Egan, Kate. "The Celebration of a 'Proper Product': Exploring the Residual Collectible through the 'Video Nasty.'" In *Residual Media: Residual Technologies and Culture*, edited by Charles R. Acland, 200–221. Minneapolis: University of Minnesota Press, 2007.

"Ekrān-e film-e '50 kilu ālbālu' motevaqqef mishavad" [Screening of "50 kilos of sour cherries" will be suspended]. *BBC Persian.* Last modified June 6, 2016. http://www.bbc.com/persian/arts/2016/06/160606_l51_50kg_sour_cherries_cinema.

Ellickson, Robert C. *Order without Law: How Neighbors Settle Disputes.* Cambridge, MA: Harvard University Press, 1994.

"Ettelāʿiyeh: Avvalin jashnvāreh-ye videʾoi-ye sureh" [Announcement: The first Sureh Video Festival]. *Sorush.* October 2, 1994.

Ezzati, Amir. *Videʾo va resāneh-hā-ye tasviri* [Video and visual media]. Bist-o-panj sāl-e sinemā-ye Iran [Twenty-five years of Iranian cinema]. Tehran: Cinema Museum, 2003.

Ezzati, Amir. "Hekāyat-e enteshār-e ketāb bedun-e nām-e moʾallef" [The story of publishing a book without the name of the author]. *Bashgāh-e adabiyyāt.* June 12, 2016. https://yadi.sk/i/n9hJEFZ7sSM3v.

Farahani, Samaneh. "Negātiv-hāʾi keh sukhtand: negāhi beh ejrā-ye tarh-e jamʿāvari-ye film-hā-ye khāreji" [The negatives that were burned: A look at the implementation of the plan to collect foreign films]. *Shahr-e Ketāb* 6 (January 2015): 70–72.

Farzin, Media. "The Video People." *Bidoun.* Accessed December 15, 2020. https://bidoun.org/articles/the-video-people.

Fathi, Nazila. *The Lonely War: One Woman's Account of the Struggle for Modern Iran.* New York: Basic Books, 2014.

Felski, Rita. *The Limits of Critique.* Chicago: University of Chicago Press, 2015.

"Film-e 'Zedd-e goluleh'" [The film "Bulletproof"]. *Bāshgāh-e film-e Tehran.* Last modified September 2, 2011. http://filmclub.vcp.ir/142558-%D9%81%DB%8C%D9%84%D9%85-%D8%B6%D8%AF-%DA%AF%D9%84%D9%88%D9%84%D9%87.

"Film-hāʾi keh dar bāzār-e videʾoi khāhim did" [What films will we see in the video market?]. *Sorush* 13. September 4, 1993.

"Filmsāzi va emkānāt-e vide'o" [Filmmaking and the possibilities of video]. *Sinemā dar vide'o*. July 1982.

Fish, Laura. "Arisen from the Grave: Collecting and Distributing Mid-century Iranian Popular Cinema." PhD diss., University of Texas at Austin, 2019.

Florida, Richard. *The Rise of the Creative Class and How It's Transforming Work, Leisure, Community and Everyday Life*. New York: Basic Books, 2002.

Foucault, Michel. *An Archeology of Knowledge*. Translated by A. Sheridan Smith. New York: Pantheon Books, 1972.

Frick, Caroline. *Saving Cinema: The Politics of Preservation*. New York: Oxford University Press, 2011.

Gabrys, Jennifer. *Digital Rubbish: A Natural History of Electronics*. Ann Arbor: University of Michigan Press, 2011.

Ganjavi, Mehdi. "A TV Show for the Rouhani Era: *Shahrzad* and the 1953 Coup d'etat in Iran." *Ajam Media Collective*. Last modified August 15, 2016. https://ajammc.com/2016/08/15/shahrzad-1953-coup-in-iran/.

Ghazian, Hossein. "The Crisis in the Iranian Film Industry and the Role of the Government." In *New Iranian Cinema: Politics, Representation and Identity*, edited by Richard Tapper, 77–85. New York: I. B. Tauris, 2002.

Gitelman, Lisa. *Already Always New: Media, History, and the Data of Culture*. Cambridge, MA: MIT Press, 2006.

Glynn, Tom. *Reading Publics: New York City's Public Libraries, 1754–1911*. New York: Fordham University Press, 2017.

Gray, Ann. *Video Playtime: The Gendering of a Leisure Technology*. London: Routledge, 1992.

Gray, Jonathan. "Mobility through Piracy, or How Steven Seagal Got to Malawi." *Popular Communication: International Journal of Media and Culture* 9, no. 2 (2011): 99–113.

Greenberg, Joshua M. *From Betamax to Blockbuster: Video Stores and the Invention of Movies on Video*. Cambridge, MA: MIT Press, 2008.

Habermas, Jürgen "The Public Sphere." *Rethinking Popular Culture: Contemporary Perspectives in Cultural Studies*, edited by Chandra Mukerji and Michael Schudson, 398–404. Berkeley: University of California Press, 1991.

Haghdad Mofrad, Ghazaleh. "The Emergence of 'Home Shows': The Market and Politics of Television Entertainment in Iran." *Ajam Media Collective*. Last modified November 14, 2014. https://ajammc.com/2014/11/17/iranian-home-shows/.

Harris, Kevan. *A Social Revolution: Politics and the Welfare State in Iran*. Berkeley: University of California Press, 2017.

Hashemi, Manata. *Coming of Age in Iran: Poverty and the Struggle for Dignity*. New York: New York University Press, 2020.

Havens, Timothy, and Amanda D. Lotz. *Understanding Media Industries*. New York: Oxford University Press, 2015.

Herbert, Daniel. *Videoland: Movie Culture at the American Video Store*. Berkeley: University of California Press, 2014.

Hilderbrand, Lucas. "Grainy Days and Mondays: Superstar and Bootleg Aesthetics." *Camera Obscura* 19, no. 3 (2004): 56–91.

Hilderbrand, Lucas. *Inherent Vice: Bootleg Histories of Videotape and Copyright*. Durham, NC: Duke University Press, 2009.

Hill, Erin. *Never Done: A History of Women's Work in Media Production*. New Brunswick, NJ: Rutgers University Press, 2016.

Hills, Matt. "Transnational Cult and/as Neoliberalism: The Liminal Economies of Anime Fansubbers." *Transnational Cinemas* 8, no. 1 (2016): 80–94.

Hirsch, Marianne, and Leo Spitzer. "Testimonial Objects: Memory, Gender, and Transmission." *Poetics Today* 27, no. 2 (Summer 2006): 353–383.

Hoek, Lotte. *Cut-Pieces: Celluloid Obscenity and Popular Cinema in Bangladesh*. New York: Columbia University Press, 2013.

Hope, Adrian. "Sistem-e kāset-e khānegi-ye zabt-e barnāmeh-hā" [The home cassette system for recording programs]. Translated by Rahim Qasemian. *Tamāshā*. June 17, 1978.

Hu, Tung-Hui. *A Prehistory of the Cloud*. Cambridge, MA: MIT Press, 2016.

Hutcheon, Linda, and Mario J. Valdes. "Irony, Nostalgia, and Postmodern: A Dialogue." *Poligraphias* 3 (2000): 18–41.

"Iranian Film Biz Revisited: Lotsa U.S. Cassettes, Picture Backlog." *Variety*, June 6, 1984.

"Jadidtarin jelveh filmfārsi dar 'Panjāh kilo ālbālu'" [The newest sparkle of filmfārsi in "50 kilos of sour cherries"]. *Cinema Press*. Last modified April 17, 2016. http://www.iranchamber.com/calendar/converter/iranian_calendar_converter.php.

Jafari, Aliakbar. "Two Tales of a City: An Exploratory Study of Cultural Consumption among Iranian Youth." *Iranian Studies* 40, no. 3 (Spring 2007): 367–383.

Jenkins, Henry. *Textual Poachers: Television Fans and Participatory Culture*. London: Routledge, 1992.

Juffer, Jane. *At Home with Pornography: Women, Sexuality, and Everyday Life*. New York: New York University Press, 1998.

Karimi, Pamela. "Alternative Belongings: Instituting and Inhabiting the Iranian Underground." In *After Belonging: The Objects, Spaces, and Territories of the Ways We*

Stay in Transit, edited by Lluís Alexandre Casanovas Blanco et al., 98–109. Zurich: Lars Müller Publishers, 2016.

Karimi, Pamela. *Domesticity and Consumer Culture in Iran: Interior Revolutions of the Modern Era*, 143–162. New York: Routledge, 2011.

Kazemi, Abbas. "Ezhdehā'i keh tarbiyyat shod: masraf-e vide'o va zendegi-ye ruz-marreh-ye daheh-ye shast" [A dragon tamed: Consuming video and daily life in the 1980s]. *Andisheh-ye Puyā* 22 (December 2014).

Khamenei, Ali. *Farhang va tahājom-e farhangi* [Culture and cultural assault]. Tehran: Sāzmān-e madārek-e farhangi-ye enqelāb-e eslāmi, 1994.

Khan, M. A. "Vidcassette Smuggling Major Headache to Pakistan Exhibs," *Variety*, May 28, 1980.

"Khatar-e vide'o barāye sinemā cheqadr jeddi ast?" [How serious is video's threat to cinema?]. *Sinemā dar vide'o.* November 1982.

Khosravi, Shahram. *Young and Defiant in Tehran*. Philadelphia: University of Pennsylvania Press, 2009.

Klinger, Barbara. *Beyond the Multiplex: Cinema, New Technologies, and the Home*. Berkeley: University of California Press, 2006.

Klinger, Barbara. "Contraband Cinema: Piracy, *Titanic*, and Central Asia." *Cinema Journal* 49, no. 2 (Winter 2010): 106–124.

Larkin, Brian. "Degraded Images, Distorted Sounds: Nigerian Video and the Infrastructure of Piracy." *Public Culture* 16, no. 2 (Spring 2004): 289–314.

Larkin, Brian. "The Poetics and Politics of Infrastructure." *Annual Review of Anthropology* 42 (2013): 327–343.

Larkin, Brian. *Signal and Noise: Media, Infrastructure, and Urban Culture in Nigeria*. Durham, NC: Duke University Press, 2008.

Latour, Bruno. *Pandora's Hope: Essays on the Reality of Science Studies*. Cambridge, MA: Harvard University Press, 1999.

Lee, Hye-Kyung. "Participatory Media Fandom: A Case Study of Anime Fansubbing." *Media, Culture & Society* 33, no. 8 (2011): 1131–1147.

Lobato, Ramon. "The Paradoxes of Piracy." In *Postcolonial Piracy: Media Distribution and Cultural Production in the Global South*, edited by Lars Eckstein and Anja Schwarz, 121–134. London: Bloomsbury Academic, 2014.

Lobato, Ramon. *Shadow Economies of Cinema: Mapping Informal Film Distribution*. London: BFI/Palgrave Macmillan, 2012.

Lobato, Ramon. "The Six Faces of Piracy: Global Media Distribution from Below." In *The Business of Entertainment. Volume 1*, edited by R. C. Sickels, 15–36. Westport, CT: Greenwood Press, 2008.

Lobato, Ramon, and Julian Thomas. *The Informal Media Economy*. Cambridge: Polity Press, 2015.

Mahdavi, Pardis. *Passionate Uprisings: Iran's Sexual Revolution*. Stanford, CA: Stanford University Press, 2009.

Mandanipour, Shahriar. *Censoring an Iranian Love Story*. Translated by Sara Khalil. New York: Vintage Books, 2009.

"Mani Haghighi: mikhāham bā kāgardāni-ye '50 kilu ālbālu' sinemā-ye tejāri rā tajrobeh konam" [By directing "50 kilos of sour cherries" I want to experiment with commercial cinema]. *Salām Sinamā*. Last modified August 29, 2015. https://www.salamcinama.ir/news/2659/.

"Mas'aleh-ye vide'o va moshkelāt va 'avārez-e farhangi-ye ān" [The issue of video and its cultural problems and duties]. Islamic Parliament Research Center of the Islamic Republic of Iran. Last modified June 30, 1992. http://rc.majlis.ir/fa/law/show/100181.

Mattelart, Tristan. "The Changing Geographies of Pirate Transnational Audiovisual Flows." *International Journal of Communication* 10 (2016): 3503–3521.

Mayer, Vicki. *Below the Line: Producers and Production Studies in the New Television Economy*. Durham, NC: Duke University Press, 2011.

McCarthy, Anna. *Ambient Television: Visual Culture and Public Space*. Durham, NC: Duke University Press, 2001.

McRobbie, Angela. *Be Creative: Making a Living in the New Culture Industries*. Cambridge: Polity Press, 2016.

Menke, Manuel. "Seeking Comfort in Past Media: Modeling Media Nostalgia as a Way of Coping with Media Change." *International Journal of Communication*, no. 11 (2017): 626–646.

Metz, Christian. *The Imaginary Signifier: Psychoanalysis and the Cinema*. Translated by Celia Britton. Bloomington: Indiana University Press, 1982.

"Mideast Entertainment Void Prompts Homevideo Bonanza." *Variety*, May 13, 1981.

Millar, Kathleen M. "The Precarious Present: Wageless Labor and Disrupted Life in Rio de Janeiro, Brazil." *Cultural Anthropology* 29, no. 1 (2014): 32–53.

Moaveni, Azadeh. *Lipstick Jihad: A Memoir of Growing up Iranian in America and American in Iran*. New York: Public Affairs, 2005.

Mohindoost, Esmaeil. *Ru dar ru bā Asghar Farhadi* [In conversation with Asghar Farhadi]. Tehran: Rownaq, 2016.

Morovvati, Saeed. "Nostālji-e posht-e mu va shalvār khomrehi" [The nostalgia behind the hair and baggy pants]. *Hamshahri*. Last modified February 5, 2019. http://newspaper.hamshahri.org/id/47172/.

"Mostafa Kiyaee: Filsuf nistam va sho'ār nemidaham/Fajr barā-ye man vitrin ast" [Mostafa Kiyaee: I am not a philosopher nor a fan of slogans/Fajr is a showcase for me]. *Islamic Azad University News Agency.* February 17, 2016: https://ana.press/fa/news/63/65872/.

Mostaghimi, Shamim. "Az sinemā beh vide'o: faryād-hā va najvā-hā" [From cinema to video: Shouts and whispers]. *Andisheh-ye puyā* 22 (December 2014).

Mottahedeh, Negar. *Displaced Allegories: Post-Revolutionary Iranian Cinema.* Durham, NC: Duke University Press, 2008.

Naficy, Hamid. "Islamizing Film Culture in Iran: A Post-Khatami Update." In *New Iranian Cinema,* edited by Richard Tapper, 26–65. London: I. B. Tauris, 2002.

Naficy, Hamid. *The Making of Exile Cultures: Iranian Television in Los Angeles.* Minneapolis: University of Minnesota Press, 1993.

Naficy, Hamid. *A Social History of Iranian Cinema, Volume 1: The Artisanal Era, 1897–1941.* Durham, NC: Duke University Press, 2011.

Naficy, Hamid. *A Social History of Iranian Cinema, Volume 2: The Industrializing Years, 1941–1978.* Durham, NC: Duke University Press, 2011.

Naficy, Hamid. *A Social History of Iranian Cinema, Volume 3: The Islamicate Period, 1978–1984.* Durham, NC: Duke University Press, 2012.

Naficy, Hamid. *A Social History of Iranian Cinema, Volume 4: The Globalizing Era, 1984–2010.* Durham, NC: Duke University Press, 2012.

Nafisi, Azar. *Reading Lolita in Tehran: A Memoir in Books.* New York: Random House, 2003.

"*Nahang-e 'anbar*: Nostālji-e daheh-ye shasti barāye ehyā-ye film-hā-ye kābārehi" [*Sperm Whale*: 1980s generation's nostalgia for the revival of cabaret films]. *Cinema Press.* Last modified December 14, 2017. http://cinemapress.ir/news/91669.

"*Nahang-e 'anbar* vāqa'an arzesh-e didār dārad" [*Sperm Whale* is truly worth seeing]. *Persianv.* Last modified June 1, 2015. http://persianv.com/cinema/%D9%81%DB%8C%D9%84%D9%85-%D9%86%D9%87%D9%86%DA%AF%D8%B9%D9%86%D8%A8%D8%B1-%D9%88%D8%A7%D9%82%D8%B9%D8%A7-%D8%A7%D8%B1%D8%B2%D8%B4-%D8%AF%DB%8C%D8%AF%D9%86-%D8%AF%D8%A7%D8%B1%D8%AF-%D8%AA%D8%B5.html.

"Naqd va barrasi-ye 'Nahang-e 'anbar-e 2: selekshen-e Roya'" [A review of "Sperm whale 2: Roya's selection"]. *Naqd-e fārsi.* Last modified June 2, 2017. http://www.naghdefarsi.com/iran-movie-review/18462-nahang-anbar.html.

"Navār-e vide'o rasman mord" [The videocassette has officially died]. *Bartarinhā.* Last modified July 23, 2016. http://www.bartarinha.ir/fa/news/357615.

"Nazar-e Mahnaz Afshar darbāreh-ye *Nahang-e 'anbar* va nostālji-e daheh-ye shast" [Mahnaz Afshar's opinion about *Sperm Whale* and the 1980s generation's nostalgia].

Nasim News Agency. Last modified August 4, 2015. http://old.nasimonline.ir /Content/Detail/1017748/.

Negus, Keith. "The Work of Cultural Intermediaries and the Enduring Distance between Production and Consumption." *Cultural Studies* 16, no. 4 (2002): 501–515.

"Nejāt-e 'Nahang-e 'anbar' bā bāzigushi-ye Attaran / Tekrār-e filmfārsi-hā-ye bi-hoviyyat" [The redemption of "Sperm whale" with Attaran's mischief / The repetition of nondescript filmfārsi movies]. *Farhang News*, July 4, 2015. http://www .farhangnews.ir/content/130946.

Newman, Michael Z. *Video Revolutions: On the History of a Medium.* New York: Columbia University Press, 2014.

Nguyen, Lilly U. "Infrastructural Action in Vietnam: Inverting the Techno-Politics of Hacking in the Global South." *New Media and Society* 18, no. 4 (2016): 637–652.

Nucho, Joanne Randa. *Everyday Sectarianism in Urban Lebanon: Infrastructure, Public Services, and Power* (Princeton: Princeton University Press, 2017).

Pahlevan, Changiz. "Negāhi beh mowqe'iyaat-e yek rasāneh-ye khatarnāk" [A look at the place of a dangerous medium]. *Māhnāmeh-ye Kelk* 1, no. 3 (May 1990): 106–110.

Pang, Laikwan. *Creativity and Its Discontents: China's Creative Industries and Intellectual Property Rights Offenses.* Durham, NC: Duke University Press, 2012.

Parks, Lisa. "Around the Antenna Tree: The Politics of Infrastructural Visibility." *Flow.* March 5, 2010. https://www.flowjournal.org/2010/03/flow-favorites -around-the-antenna-tree-the-politics-of-infrastructural-visibilitylisa-parks-uc -santa-barbara/.

Parks, Lisa. "Media Infrastructures and Affect." *Flow.* May 19, 2014. https://www .flowjournal.org/2014/05/media-infrastructures-and-affect/.

Parks, Lisa. "'Stuff You Can Kick': Toward a Theory of Media Infrastructures." In *Between Humanities and the Digital*, edited by Patrik Svensson and David Theo Goldberg, 355–373. Cambridge, MA: MIT Press, 2015.

Parks, Lisa, and Nicole Starosielski. "Introduction." In *Signal Traffic: Critical Studies of Media Infrastructures*, edited by Lisa Parks and Nicole Starosielski, 1–28. Urbana: University of Illinois Press, 2015.

Parsa, F. "Letter from Tehran." Translated by Dorna Khazeni. *Film-Philosophy* 7, no 12 (June 2003). http://www.film-philosophy.com/vol7-2003/n12parsa.

Partovi, Pedram. *Popular Iranian Cinema before the Revolution: Family and Nation in Filmfārsi.* New York: Routledge, 2017.

Perren, Alissa. "Rethinking Distribution for the Future of Media Industry Studies." *Cinema Journal* 50, no. 3 (Spring 2013): 165–171.

Pertierra, Anna Cristina. "Private Pleasures: Watching Videos in Post-Soviet Cuba." *International Journal of Cultural* Studies 12, no. 2 (2009): 113–130.

"Porforushtarin film-hā-ye sinemā-ye Iran dar sāl-e 1391 hamrāh bā 'aks" [The best-selling movies in Iranian cinema in 1391 along with pictures]. *Bizhan Zadmehr.* Last modified February 24, 2013. http://bijanzadmehr.blogfa.com/post /59/.

"Porsesh az dāneshjuyān: Cherā pā-ye vide'o mineshinid va az ān cheh estefādeh-hā'i mikonid?" [Asking university students: Why do you watch video and what do you do with it?]. *Gardun,* September–October 1993.

Portelli, Alessandro. *The Death of Luigi Trastulli and Other Stories: Form and Meaning in Oral History.* Albany: SUNY Press, 1991.

Qaderi, Kaveh. "Ghelzat-e Nostāljik va portreh-ye nimeh tamām" [Nostalgic consistency and an incomplete portrait]. *Cinema Screen.* Last modified June 2015. http://www.cinscreen.com/?id=6279.

"Qānun-e nahveh-ye residegi beh takhallofāt va mojāzāt-e forushandegān-e lebās-hāi keh estefādeh az ānhā dar mala'-e 'ām khalāf-e shar' ast va yā 'effat-e 'omumi rā jarihehdār mikonad" [Law dealing with the violations and punishments of vendors of clothing whose use in public spaces is against religious law or violates public chastity]. Islamic Parliament Research Center of the Islamic Republic of Iran. Last modified March 18, 1987. http://rc.majlis.ir/fa/law/show/91340.

Rancière, Jacques. *Proletarian Nights: The Workers' Dream in Nineteenth-Century France.* Translated by John Drury. New York: Verso, 2012.

Rekabtalaei, Golbarg. *Iranian Cosmopolitanism: A Cinematic History.* Cambridge: Cambridge University Press, 2019.

Romney, Jonathan. "Taxi Tehran Review—Jafar Panahi's Joyous Ride." *The Guardian.* Last modified November 1, 2015. https://www.theguardian.com/film /2015/nov/01/taxi-tehran-review-admirable-jafar-panahi-takes-to-the-streets.

Saffarian, Nasser. "Dar hāshiyeh-ye bāzdāsht-e nevisandegān-e sinemāi" [On the margins of the arrests of the cinematic writers]. Last modified April 9, 2003. http://www.saffarian.ws/Articles.aspx?ID=fd5e907510be4701b668487e0c19d557.

Salamati, Maha. "Transnational Film Circulation in the Iranian Context: From Conjunctural Crisis to Discursive Heterotopia." PhD diss., University of New South Wales, 2019.

"Sargozasht-e vide'o dar Iran, qesmat-e avval" [The tale of video in Iran, part 1]. *Sorush.* August 21, 1993.

"Sargozasht-e vide'o dar Iran, qesmat-e dovom" [The tale of video in Iran, part 2]. *Sorush.* September 4, 1993.

Satrapi, Marjane. *The Complete Persepolis.* New York: Pantheon, 2000.

Scheible, Jeff. "Video after Video Stores." *Canadian Journal of Film Studies* 23, no. 1 (Spring 2014): 51–73.

Scott, James C. *Domination and the Arts of Resistance*. New Haven, CT: Yale University Press, 1990.

Scott, James C. *Weapons of the Weak: Everyday Forms of Peasant Resistance*. New Haven, CT: Yale University Press, 1985.

"Sedā-ye to khub ast: behtarin-hā-ye tārikh-e dubleh Iran" [Your voice is good: The best in the history of dubbing in Iran]. *Ketāb-e sāl-e sinemā-ye Iran, 1384* [The yearbook of Iranian cinema, 2005/2006] 15 (February 2006): 110–120.

Shahabi, Mahmood. "The Iranian Moral Panic over Video: A Brief History and a Policy Analysis." In *Media, Culture and Society in Iran: Living with Globalization and the Islamic State*, edited by Mehdi Semati, 127–145. London: Routledge, 2007.

Shahrokni, Nazanin. *Women in Place: The Politics of Gender Segregation in Iran*. Berkeley: University of California Press, 2019.

Sheibani, Khatereh. "Shahrzad: Farhang-e mardomi, farhang-e rasānehi va bāznamā'i-ye jensiyyat dar 'asr-e dijitāl" [*Shahrzad*: Pop culture, media culture, and the representation of gender in the digital age]. *Iran Namag: A Quarterly of Iranian Studies* 2, no. 2 (Summer 2017): 62–111.

Siamdoust, Nahid. *Soundtrack of the Revolution: The Politics of Music in Iran*. Stanford, CA: Stanford University Press, 2017.

Siavoshi, Sussan. "Cultural Policies and the Islamic Republic: Cinema and Book Publication." *International Journal for Middle East Studies* 28, no. 4 (November 1997): 509–530.

Simone, AbdouMaliq. "People as Infrastructure: Fragments in Johannesburg." *Public Culture* 16, no. 3 (Fall 2004): 407–429.

"Siyāsat-e pardākht-e sahmiyeh beh tahiyeh konandegān-e film-hā-ye vide'oi e'lām shod" [A policy subsidizing video movie producers was announced]. *Ettelā'āt* (Tehran). October 4, 1993.

"Sokhanān-e vazir-e ershād-e eslāmi dar mored-e lāyeh-ye matbu'āt, tablightāt-e khārej az keshvar, taklif-e honarmandān, turism, tarh-e ettelā'āti-ye keshvar va taghirāt dar vezārat-e ershād." [Comments by the minister of culture and Islamic guidance about the press, advertising outside of the country, the work of artists, tourism, the role of information in the country, and changes in the ministry of Islamic guidance]. *Ettelā'āt*. December 12, 1982.

Solhju, Tahmasb. "Khub, bad, mofid, mozer, zesht, zibā, sāzandeh, virāngar" [Good, bad, useful, harmful, ugly, beautiful, constructive, destructive]. *Film* 157 (April 1994): 97–99.

Sontag, Susan. "The Decay of Cinema." *New York Times*. February 25, 1996.

Spigel, Lynn. *Make Room for TV: Television and the Family Ideal in Postwar America*. Chicago: University of Chicago Press, 1992.

Sreberny-Mohammadi, Annabelle, and Ali Mohammadi. *Small Media, Big Revolution: Communication, Culture, and the Iranian Revolution*. Minneapolis: University of Minnesota Press, 1994.

Standage, Tom. *The Victorian Internet: The Remarkable Story of the Telegraph and the Nineteenth Century's On-line Pioneers*. 2nd ed. New York: Bloomsbury USA, 2014.

Star, Susan Leigh. "Ethnography of Infrastructure." *American Behavioral Scientist* 43, no. 3 (1999): 377–391.

Starosielski, Nicole. *The Undersea Network*. Durham, NC: Duke University Press, 2015.

Steirer, Gregory. "The Personal Media Collection in an Era of Connected Viewing." In *Connected Viewing: Selling, Streaming and Sharing Media in the Digital Age*, edited by Jennifer Hold and Kevin Sanson, 79–96. New York: Routledge, 2014.

Stevens, Dana. "The Cab Ride as Artistic Rebellion." *Slate*, October 2, 2015. https://slate.com/culture/2015/10/jafar-panahis-film-taxi-reviewed.html.

Sundaram, Ravi. *Pirate Modernity: Delhi's Media Urbanism*. London: Routledge, 2010.

Tabarraee, Babak. "*Gone with the Wind* in Iran: The Iranian Dub Industry and the Re-authorization of Classical Holllywood Cinema." YouTube video, 39:38. January 18, 2018. https://www.youtube.com/watch?v=j7bA7_KCWXs.

Tabarraee, Babak. *Kālt* [Cult]. Tehran: Enteshārāt-e Qoqnus, 2016.

Tabarraee, Babak. "Rationalizing the Irrational: Reza Attaran's Popularity, Stardom, and the Recent Cycle of Iranian Absurd Films." *Iranian Studies* 51, no. 4 (2018): 613–632.

"Tahājom-e Farhangi va ʿazm-e melli [Cultural assault and the will of a nation]." *Film*. January 10, 1992.

Tahmuresi, Sina. "Negāhi beh film-e 'Nahang-e ʿ anbar-e 2'" [A look at "Sperm whale 2"]. *Zoomg*. Last modified May 28, 2017. http://www.zoomg.ir/2017/5/21/161561/movie-nahang-e-anbar-introduction/.

"Tārikhcheh-ye videʾo va navār-e videʾo" [The history of video and the videocassette]. *Sinemā dar videʾo* (Tehran). Tir 1361/July 1982.

Tashiro, Charles. "The Contradictions of Video Collecting." *Film Quarterly* 50, no. 2 (Winter 1996–1997): 11–18.

"Tasvir-e ebtezāl mahv mishavad" [The image of obscenity fades away]. *Sorush*. June 4, 1983.

Thomas, Mark. "Homevideo Piracy, Penetration Killing Middle East Cinema." *Variety*, May 25, 1988.

Thomas, Mark. "Iranians Just Want to Have Fun." *Variety*, February 14, 1990.

Vaidhyanathan, Siva. *Copyrights and Copywrongs: The Rise of Intellectual Property and How It Threatens Creativity*. New York: New York University Press, 2001.

Varzi, Roxanne. *Warring Souls: Youth, Media, and Martyrdom in Post-Revolution Iran*. Durham, NC: Duke University Press, 2006.

"Vide'o: Az zendegi-ye makhfi tā hozur-e āshkār" [Video: From a hidden life to an open present]. *Vide'o-māhvāreh* (March 1994).

"Vide'o va ma'muriyat-e ān dar Iran" [Video and its mission in Iran]. *Sorush*. May 31, 1986.

"Vorud-e navār-e khām-e vide'o mamnu' shod" [The entry of blank tapes has been banned]. *Film* (March 1986).

Wang, Shujen. *Framing Piracy: Globalization and Film Distribution in Greater China*. Lanham, MD: Rowman & Littlefield, 2003.

Wasser, Frederick. *Veni, Vidi, Video: The Hollywood Empire and the VCR*. Austin: University of Texas Press, 2002.

Williams, Linda. *Hard Core: Power, Pleasure, and the "Frenzy of the Visible."* Berkeley: University of California Press, 1989.

Williams, Linda. *Screening Sex*. Durham, NC: Duke University Press, 2008.

Williams, Rosalind. *Notes on the Underground: An Essay on Technology, Society, and the Imagination*. Cambridge, MA: MIT Press, 2008.

Young, Deborah. "Iran Unveils New Film Threat." *Variety*, February 17, 1992.

Zeydabadi-Nejad, Saeed. *The Politics of Iranian Cinema: Film and Society in the Islamic Republic*. London: Routledge, 2009.

INDEX

Note: Page numbers in italics indicate illustrations.

INFRASTRUCTURES SERIES

edited by Geoffrey C. Bowker and Paul N. Edwards

Megan Finn, *Documenting Aftermath: Information Infrastructures in the Wake of Disasters*

Ann M. Pendleton-Jullian and John Seely Brown, *Design Unbound: Designing for Emergence in a White Water World*, Volume 1: *Designing for Emergence*

Ann M. Pendleton-Jullian and John Seely Brown, *Design Unbound: Designing for Emergence in a White Water World*, Volume 2: *Ecologies of Change*

Jordan Frith, *A Billion Little Pieces: RFID and Infrastructures of Identification*

Morgan G. Ames, *The Charisma Machine: The Life, Death, and Legacy of One Laptop per Child*

Ryan Ellis, *Letters, Power Lines, and Other Dangerous Things: The Politics of Infrastructure Security*

Mario Biagioli and Alexandra Lippman, eds, *Gaming the Metrics: Misconduct and Manipulation in Academic Research*

Malcolm McCullough, *Downtime on the Microgrid: Architecture, Electricity, and Smart City Islands*

Emmanuel Didier, translated by Priya Vari Sen, *America by the Numbers: Quantification, Democracy, and the Birth of National Statistics*

Andrés Luque-Ayala and Simon Marvin, *Urban Operating Systems: Producing the Computational City*

Michael Truscello, *Infrastructural Brutalism: Art and the Necropolitics of Infrastructure*

Christopher R. Henke and Benjamin Sims, *Repairing Infrastructures: The Maintenance of Materiality and Power*

Stefan Höhne, *New York City Subway: The Invention of the Urban Passenger*

Timothy Moss, *Conduits of Berlin: Remaking the City through Infrastructure, 1920–2020*

Claude Rosental, translated by Catherine Porter, *The Demonstration Society*

Blake Atwood, *Underground: The Secret Life of Videocassettes in Iran*

Huub Dijstelbloem, *Borders as Infrastructure: The Technopolitics of Border Control*

Dylan Mulvin, *Proxies: Standards and Their Media*